HERITAGE FROM THE WILD
Familiar Land & Sea Mammals of the Northwest

by M. Douglas Scott and Suvi A. Scott

Number 2

NORTHWEST GEOGRAPHER™ Series
John A. Alwin, Ph.D.
Editor & Publisher

NORTHWEST GEOGRAPHER™ SERIES

John A. Alwin, Ph.D.
Editor & Publisher

Northwest Panorama Publishing, Inc.
Box 1858   Bozeman, MT 59771

Library of Congress Catalog Card Number: 85-062284

ISBN 0-9613787-1-9

Production Credits

Design and Layout: Ann Alwin, B.F.A.
                   Bozeman, Montana

Cartography: Chad Groth
             Bozeman, Montana

Typesetting: Color World Printers
             Bozeman, Montana

Printed in Japan by Dai Nippon Printing Co.,
Ltd., Tokyo

The Cover

Four-day-old black-tailed deer fawn finds shelter under alpine fir in western Washington's Olympic Mountains. Thomas W. Kitchin photo.

Title page

Steller's sea lions sun-bathe in Puget Sound. Mt. Baker looms in the distance. Ken Balcomb, III photo.

Contents page

Nature's symmetry, courtesy of Rocky Mountain elk. Richard E. Kirchner photo.

About This Series

For its size, the Northwest is one of the world's most diverse regions. Few can rival its kaleidoscope of history, natural landscapes, climates, geology, urban centers, economic activity, and peoples. A geographer couldn't ask for a more interesting study area.

Each profusely illustrated and highly readable *Northwest Geographer*™ will focus on one component region or a specific Northwest topic. This captivating series is designed for the geographer in each of us and especially for the justifiably proud residents fortunate enough to call this special place home.

To be added to the *Northwest Geographer*™ Series mailing list and receive information on prepublication discounts on future books, write:

Northwest Panorama Publishing, Inc.
NORTHWEST GEOGRAPHER Series
P.O. Box 1858
Bozeman, MT 59771

FOREWORD

Since most of us are interested in our surroundings, we all can claim to be budding geographers, college degree or not. The basic abstract unit geographers like to work with is the region. This is a place that has relative uniformity in certain major characteristics, such as climate, primary river drainages, history of settlement or even socio-economic development.

The northwestern corner of the lower 48 states is just such a place. Weather, from soggy, leaden-cloud fronts to dry, winter-relieving chinooks are gifts of the Pacific. The vast Columbia River network drains most of this 600-by-500 mile area. When the world had an insatiable demand for rich furs, the Northwest was the great provider, and the journals of early trappers and explorers like Ogden, Work, Thompson and Lewis and Clark recorded one early history for the region as a whole.

The kinds of wild mammals that provided income and food for the Northwest's earliest chroniclers are still with us today. Whereas a New England or Midwestern geographer series might easily omit wild mammals when describing those regions, this would be unthinkable in a Northwest series. Northwesterners are people who regularly watch, photograph, hunt, trap and conserve a great variety of wild mammals. The mountain goat, grizzly bear and Roosevelt elk are as much a part of our Northwestern heritage as is the Columbia River.

We hope that this volume on wild mammals of the Northwest will help to affirm and preserve our regional identity.

M. Douglas and Suvi Scott
Bozeman, MT

CONTENTS

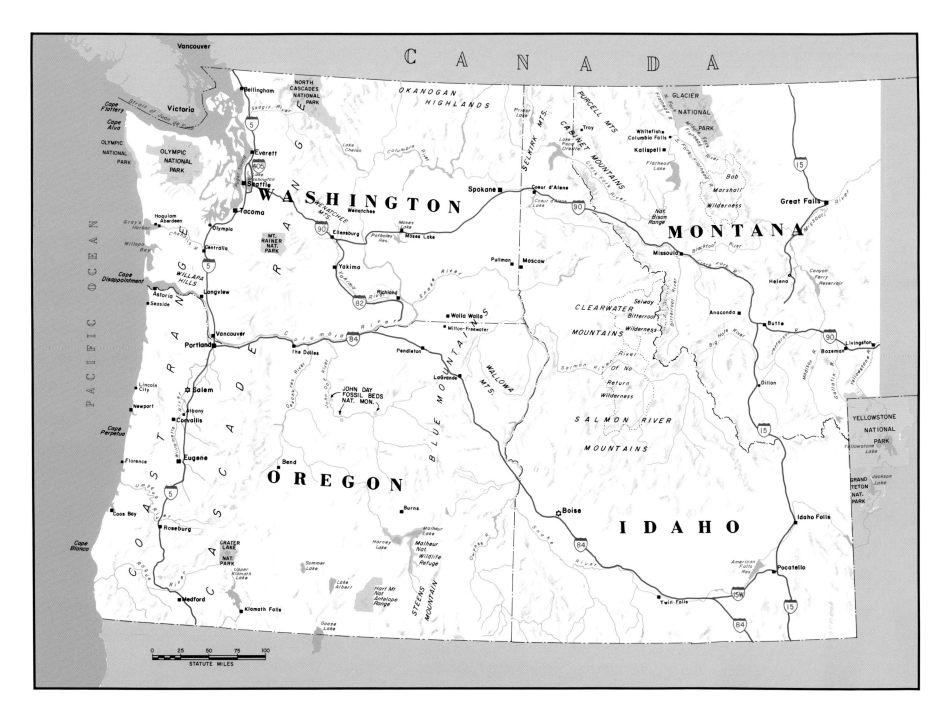

4

Introduction

Our Northwest includes the entire states of Washington, Oregon and Idaho, as well as Montana's mountainous western third. Without doubt, this region is home to the lower 48's greatest number and variety of easily identifiable wild mammals. Spotting for mountain goats in North Cascades National Park, whale watching off Oregon's Cape Perpetua, counting antelope while driving State Highway 28 between Idaho Falls and Salmon or catching a glimpse of an elusive grizzly in Montana's Bob Marshall Wilderness area are all special privileges afforded Northwest residents.

Why this area has been a refuge for such mammals as the cougar and grizzly is clear to those who know the region's physical geography. The rugged Rockies and Cascades, the semiarid prairies between the mountains and the temperate rain forests of the Olympics offered formidable obstacles to our forefathers. There were ephemeral miners, tough lumberjacks and persevering farmers who settled the Northwest, but most of the land simply could not support the same density of rural populace as the Midwest or East. Valley fences were far apart and mountains usually were places to visit, not inhabit. In some localities the larger mammals with food or fur value were over exploited and habitat was converted to human use, but there always were remote spots where the animals could live as they always had.

Time, also, was on the side of our mammals, since most of the Northwest was settled relatively late from a national perspective. Initially, growth was slow in this remote northwest corner. The pace quickened only this century with regional population ballooning from just over one million to more than eight million, many arriving since World War II. Today our lives are centered around silicon chips, rather than the buffalo kind. Metropolitan Seattle, Portland, Boise and even Mis-

Red squirrel fattens up for a forthcoming Northwest winter. Bruce Pitcher photo.

soula are booming with newcomers. Had this urban invasion occurred one or two hundred years ago when American values were far different, we probably would have said farewell to many of our most distinctive wild mammal species just as people did in most other American regions. Fortunately, before great droves of people reached the Northwest, America experienced a revolution in attitudes toward the natural environment. National concern with pollution, wilderness and endangered species is scarcely 25 years old. The Northwest's heightened environmental awareness offers great promise in helping preserve our wild mammals even as the region continues to grow.

Our primary purpose in preparing this short work is to build on people's already substantial interest in the natural world. Paraphrasing the old adage, "to know them is to like them," we hope that by discussing our familiar wild mammals we will encourage their conservation and wise management. This book is not another field guide to area mammals. Its approach is unique. While the life history of mammals is included, efforts have been made to place these Northwest residents in their regional, historical and human context. Mammals included generally are those that can be seen easily in their native habitats. Most are rabbit-size or larger and have obvious economic importance. A few less common high-interest or endangered species also are included. Our Northwest mammals are not merely curiosity items, but are integral elements in the regional mosaic of man and nature—a valuable heritage we must preserve.

Mammals, Animals, and Such

All too often we read a newspaper article, or even a fish and game department report referring to the "birds and animals" of a region. When professionals, who should know better, use such language we are tempted to enroll them in a remedial high school biology course! What people usually mean when they refer to "animals" is *mammals*. Birds, fish, insects, reptiles and mammals are all different kinds of animals.

One characteristic often associated with mammals is the presence of hair. Although hair-like structures occur on other animals (like the chest "beard" in the male turkey), true hair is found only on mammals. Many, such as the beaver and fox, have thick pelts consisting of fine, short underfur and stiff, longer guard hairs providing much of their distinctive coloration. A few mammals, including the elephant and man, have a very meager hair coat. Other mammals, whales and dolphins for instance, have no hair or just a few remnant bristles.

Hair, of course, is not just for adornment of a mammal's body. Particularly for mammals in cold regions, hair and fur provide insulation against the elements. When hair is not present, a generous layer of fat or thick skin may help insulate an animal. Insulation, plus physiological mechanisms, including shivering and increased metabolic rate, all help most mammals maintain a relatively constant body temperature day and night. In hot weather, mammals cool themselves by such means as panting or sweating.

The mammary gland, actually a modified sweat gland, is one structure all mammals share. The liquid it produces (milk) is high in minerals, fats and proteins and is used to feed all newborn mammals, giving them a quick start in life. No other animals have mammary glands, although some produce a milk-like substance from different glands.

No movie-goer who has seen "Jaws" is likely to forget that fish have teeth, as do many other animals, but mammals have turned the production of teeth into a fine art. Whereas the teeth of most animals are relatively simple and for a single purpose, many mammalian teeth have evolved into complex shapes for a number of uses. The jaws of a mammal that eats a variety of plant and animal foods may contain different teeth for nipping, piercing, shearing, crushing and grinding.

At this point mammals dominate the earth, as opposed to dinosaurs or fishes in earlier times. Biologists think one reason for this turn of events is that mammals as a group have developed relatively larger and more complex brains than any other animal group. Mammals excel at storing information in the brain and learning from experience. These abilities are highly useful in allowing mammals to cope with everyday changes in the natural environment by modifying their behavior.

Above: The Northwest's most abundant resident sea mammal, the harbor seal, may be seen at the mouth of rivers and in shallow harbors. Thomas L. Spaulding photo.

Left: Like the young of other mammal species, this bison calf depends on its mother's milk for a quick start in life. Bruce Pitcher photo.

Solo performance by Canis latrans, *the song dog of the Northwest. Bruce Pitcher photo.*

What's In a Name?

Sooner or later, anyone who reads much about living things will discover that every plant and animal has a somewhat mysterious scientific name along with its much more understandable common name. People seem to have a propensity for giving a label to everything around them, and familiar plants and animals probably have had common names since the first primitive humans conversed. These simple names for living things satisfied people's needs until early naturalists realized that some standardized method of classifying and identifying organisms was required if any sense

was to be made out of the thousands of new forms that were being discovered by advancing civilization.

From the time of Aristotle onward, many systems of categorizing and naming living things were developed and discarded. This effort took a great leap forward during the 18th century when Swedish botanist, Carolus Linnaeus, popularized the use of two separate Latin terms, called binomial nomenclature, to identify each living thing. The first part of a binomial scientific name is called the genus (plural genera), and the second part is the species name. The genus

name always is capitalized, while the species name usually is not. As an example, the scientific name for the familiar coyote is *Canis latrans.* The genus name can be used for several closely related species, but cannot be used with any other group in the animal kingdom. Thus, the closely related dog (*Canis familiaris*) and wolf (*Canis lupus*) are both in the genus *Canis.* The species name, however, may be used for only one kind of animal in a given genus, but it may be used again in another genus. For instance, we have *Castor canadensis* (the beaver) and *Ovis canadensis* (the bighorn sheep). The result of these two rules is that every living thing has its own unique set of names. In many cases, biologists also have discovered local varieties of mammal species, and these are given a third scientific name, which is the subspecies name.

One purpose of scientific nomenclature is to avoid the confusion caused by use of local common names. For example, Europeans call *Alces alces* an elk, while Northwesterners call the same animal a moose, and our elk (*Cervus elaphus*) is something quite different. Confusion of common names does not result just from living in different countries. People in the Northwest commonly call the Townsend's ground squirrel (*Spermophilus townsendii*) a gopher, while Midwesterners use the same common name for the northern pocket gopher (*Thomomys talpoides*), which we see from the genus name is not even closely related.

With a little knowledge of Latin or Greek word roots, we can learn something about the animal from the scientific name. Using our coyote friend as an example, *Canis* comes from the Latin word for "a dog," and *latrans* is derived from a Latin term meaning "barking." Those familiar with the vociferous habits of this doglike animal readily would agree the scientific name is apt.

Hoofed Mammals

Bighorn sheep. Bruce Pitcher photo.

Rocky Mountain Elk and Roosevelt Elk

Elk inhabiting Washington's Olympic Peninsula got lucky in 1897. Naturalist C. Hart Merriam named the race to which they belong, *roosevelti*, to honor the conservationist who became United States President in just four years. A change in fortune was badly needed, since herd numbers had dropped drastically during the latter part of the 19th century due to over hunting and habitat destruction.

In 1903 the Elks Lodge in Port Angeles asked Congress to create a national park on the Olympic Peninsula for protection of the elk. A subsequent bill failed to pass, but another, more direct course of action was available. Washington's Congressman William E. Humphrey approached Theodore Roosevelt in early 1909 about creating a game reserve in the Olympic Mountains. Possibly, in part, because his namesake was involved, Roosevelt was sympathetic to the cause, and quickly signed a Presidential Proclamation creating Mount Olympus National Monument. The Roosevelt elk were protected, and the seeds for Olympic National Park were firmly planted.

The generic common name, elk, has obscure origins in Europe, where it is applied to the Old World moose. Because of this confusion, many biologists prefer the common name, wapiti, a Shawnee Indian name for the American elk. Roosevelt elk originally inhabited Washington and Oregon west of the Cascades. Presently only the coastal portions of these two states have pure strains, with the largest herd of 5,000 residing within Olympic National Park. In the early 1900s, Rocky Mountain elk were transplanted from Yellowstone Park into northwestern Washington, where they freely interbred with the native subspecies, creating the hybrids encountered in the North Cascades today. Descendants of 21 Rocky Mountain elk stocked in Washington's central Cascades in the early 20th century today form the Yakima herd of some 10,000 animals. The nearby herd in the Wenatchee Mountains north of Ellens-

Rocky Mountain elk depicted by John James Audubon in: John J. Audubon and the Reverend John Bachman, Quadrupeds of North America *(3 vols.; New York: V.G. Audubon, [1849]-1860). Unless otherwise indicated, all lithographs that follow were drawn by J.J. Audubon or his son J.W. Audubon and appear in this work, cited as Audubon,* Quadrupeds. *Photo from Special Collections Division, University of Washington Libraries.*

burg, many elk ranging in northeastern Oregon, and those in the western part of the state outside coastal areas, as well as other herds in Washington and Oregon are descendants of animals stocked from Yellowstone. Today, due to transplant efforts, these ungulates occupy nearly all suitable habitat in these two states, and their numbers are thought to be higher than in 1800. Rocky Mountain elk also are found throughout the mountainous regions of Idaho and western Montana.

Roosevelt elk always have preferred the heavily timbered mountainous areas typical of western Oregon and Washington, whereas their Rocky Mountain relatives formerly lived in both the plains and the mountains. Settlement of major lowlands in the Northwest largely eliminated the usefulness of this habitat, which may have been especially valued as winter range. Contrary to popular opinion, Rocky Mountain elk are not plains animals that have been driven to the mountains by encroaching civilization. Even Lewis and Clark noted that elk were more numerous in the forests than on the plains.

Most Rocky Mountain elk (*Cervus elaphus*) spend summers grazing in alpine meadows adjacent to semi-open forests. They often retreat to lower and denser coniferous cover, and small grass-covered valleys for the snowy season. Some elk move up to 50 miles between summer and winter ranges, while others are basically nonmigratory. A good example is an elk population living along Idaho's Selway River. Part of the individuals migrate long distances during spring and fall, whereas others make only minor shifts in habitat between seasons.

Elk are mainly nocturnal, though they do much of their foraging around sunrise and dusk. Their diets consist of more than 300 plant species. In spring, many kinds of grasses and sedges are eaten. During summer, forbs (including wild onion and dandelion) constitute the bulk of the diet, while succulent leaves of shrubs also are relished. The fall diet shifts back to grasses and sedges. In winter the importance of browse is pronounced in some areas of the Northwest. Where elk winter in grasslands, like in portions of western Montana, 65 to 100 percent of their diet consists of grasses.

If a winter is abnormally severe, wapiti may raid haystacks and other agricultural products. The exceptionally heavy snowfalls of 1985 in northeastern Oregon forced elk to seek forage at unusually low elevations. As a result they damaged over 100 acres of apple orchards along the north and south forks of the Walla Walla River near Milton-Freewater. Elk damage to crops and fences in the Baker, Oregon area have been so severe that the Oregon Department of Fish and Wildlife initiated a feeding program, trying to keep the animals in the mountains. Overlap between elk winter range and subdividable agricultural land has caused headaches for land-use planners in Baker County, who are required by state law to provide for wildlife when developing comprehensive land-use plans.

Elk are social creatures. As soon as a newborn calf is a fortnight old it accompanies its mother in joining a nursery herd of other cows and calves. Such groups graze within summer home ranges encompassing up to two square miles. As the breeding season nears in autumn, the otherwise solitary bulls establish territories, which they mark by shredding bark off small trees with their antlers. In heavily hunted populations, like those

Above: Rocky Mountain elk cow. Bruce Pitcher photo.

Left: Roosevelt elk on the Columbia White-tailed Deer National Wildlife Refuge near Cathlamet, Washington. Thomas W. Kitchin photo.

in the Blue Mountains of Oregon, yearling bulls may do most of the breeding, while in more inaccessible areas it is primarily done by males at least three years old. At the onset of the rut in early September, the bulls announce their presence and challenge rivals with a loud, eerie, squealing call—the bugle. They challenge a nearby adversary by thrashing bushes and digging dirt with their antlers. If such threat displays fail to convince the opponent to retreat, a battle ensues with combatants locking antlers and shoving each other. Encounters are seldom fatal, although the broken tines so common in a big bull's antlers testify to the intensity of the jousts.

Bull elk begin growing antlers in May. Antlers are not properly called horns, for they differ from most true horns in two major ways. They are formed of bone-like material, not hair-like material, and secondly, they are shed annually. While antlers develop they are covered by skin, under which numerous blood vessels provide necessary nourishment for the rapidly growing bone. Nerves also are present, and the antlers are sensitive to touch. Fine grayish-brown hair, which looks and feels like velvet, coats the skin surface. When antlers attain their full size by early August, the velvet dries out and falls off, a process hastened by the animal when it rubs the antlers against trees and shrubs while marking a territory. Yearling male elk, like most deer, have only spikes as antlers, but the older animals sport branched headsets. Antlers reach their greatest size in mature bulls over five years of age, who can boast racks weighing 30 pounds or more, measuring up to five feet in length, and having most typically six points along each beam. Antlers of Roosevelt elk are shorter and thicker with crowned, or webbed, tines at the top instead of the longer, straighter tines of their Rocky Mountain counterparts.

Bull elk spend about as much energy growing their antlers as do females in producing their calves. The antlers usually are shed in late March or early April. People in the Northwest have found that Orientals will pay $6.00 a pound or more for freshly dropped elk antlers. Elk ranches sell surgically removed antlers in velvet for $100 or more a pound! Buyers sell the antlers to be ground and blended into various concoctions that have supposed medicinal and aphrodisiacal qualities. The

Left: Two Montana bull elk tangle. Such encounters seldom are fatal. Richard E. Kirchner photo.

Below: Antlers locked, sparring partners have at it. Alan Carey photo.

final product may command a price of $50 or more per *ounce*.

Although commonly believed, a bull does not gather a harem, rather it seeks out and joins one of the existing cow-calf groups. After declaring himself the herd bull he may have to fight off several contenders who also have amorous intentions. By late November the rut will come to an end. Most females over two breed annually. When forage is scarce, cows commonly breed only in alternate years, which often occurs among Oregon's Roosevelt elk.

Wapiti seldom live more than 20 years. Generally, life expectancy for males is shorter than females, because of the depletion of fat reserves during the rut. Malnourished bulls sometimes cannot regain their strength on low-energy browse available in winter and may succumb before the arrival of spring.

A single calf normally is born in late May or early June, following a nine-month gestation period. The traditional calving grounds are usually small grassy openings in wooded areas.

Calves arrive well camouflaged with their light-colored spots against a russet pelage. At birth the young weigh about 32 pounds. They rely entirely on mother's milk for a month.

Wapiti bulls weigh between 600 and 900 pounds with an occasional giant reaching more than 1,000 pounds. Adult cows weigh 400 to 600 pounds. Bull Rocky Mountain elk average nearly eight feet in length and four to five feet in height while Roosevelt elk are slightly larger. During summer the Rocky Mountain variety are a sleek brownish-red. The fall molt of hair changes this to tan pelage, with a dark brown head and neck. Roosevelt elk are generally a darker brown than their cousins in all seasons.

Natural enemies of the wapiti are scarce. Mountain lions and wolves may take some adults during winter, while grizzlies sometimes attempt to capture those weakened by winter's inadequate diet. In some places, like the Lochsa River area of Idaho, black bears are known to have caused significant losses among calves less than a month old.

Below left: Hours-old Rocky Mountain elk calf tests its legs and surveys its new world. Richard E. Kirchner photo Above left: Alert young members of a nursery herd of Rocky Mountain elk sense an intruder. Bruce Pitcher photo Above: Procession of wapiti bulls punctuates the horizon. Bruce Pitcher photo Right: Northwest monarch wears a crown of velvet. Richard E. Kirchner photo. Despite the animals' association with forested mountain environments, a herd of 55 elk lives year-round on the desert flats of central Washington's Hanford Nuclear Reservation.

Carpet of wildflowers provides a pristine final resting place. Richard E. Kirchner photo.

It has been said that a great bull elk is the finest game animal. The enormous antlers are among the most impressive in the world, and the big deer provides several hundred pounds of some of the finest meat there is. When hunted in mountainous areas, a wapiti bull frequently seeks out the most miserable, snowy, steep, downfall-choked ravines it can find. Unless a hunter is extremely lucky it is a very difficult species to find and stalk. Elk hunter success ranges from about 10 percent in Washington to as high as 25 percent in Idaho—chances of taking home a big bull elk are significantly less.

In spite of the difficulties, persistent hunters bag 50,000 to 60,000 elk in the Northwest each year. Regular readers of *Outdoor Life* and *Field and Stream* magazines know that western Montana and northern Idaho have world-class reputa-

tions for elk hunting. Some out-of-state hunters willingly pay outfitters $3,000 and more for a one-week hunt into the northern Rockies backcountry. Such high-quality hunts in the infamous Bob Marshall Wilderness and other legendary haunts contrast markedly with some recent hunts of central Washington's Colockum herd. If snowfall is absent or light, the herd within the Colockum Management Area is easily accessible by car. This guarantees the thousands of hunters who storm this small eastern corner of the Wenatchee Mountains. Hunter density even has justified temporary coffee shops at the busiest dirt-road intersections. More than once, extreme hunter pressure has forced fleeing animals to swim across the nearby Columbia River!

White-Tailed Deer and Columbian White-Tailed Deer

Just about any crop a human wants to grow, a white-tailed deer is willing to eat. Garden vegetables, emerging cereal grains, apples, seed alfalfa, field corn, petunias, golf course greens, shrubbery, haystacks and young conifers all make their way to this animal's accommodating paunch. Many Northwest hunters have found ranchers happy to have them use a haystack for a deer blind, but who would think of hiding behind a Christmas tree? The opportunity may arise in northwestern Montana's Flathead Valley. Bigfork-area Christmas tree growers claim they lost a quarter- to a half-million-dollars worth of young trees during the winter of 1984-85 as a result of hungry white-tailed deer clipping off the tops. Problems like this usually occur only when deer populations are abnormally high. More often, losses are negligible and most ranchers and gardeners enjoy the novelty of having a few deer in their backyards.

White-tails (*Odocoileus virginianus*) do most of their foraging at twilight and in the early morning hours before sunrise, spending their days resting and ruminating within a secure woody patch. These deer are known to feed on more than 1,000 species of plants, although only about 100 are truly relished. They have an amazing ability to consistently select foods with the highest nutritional value. During the cold season in the Northwest the animals' diet usually centers around twigs and buds of browse species, including conifers. In spring and summer succulent leaves and stems of various forbs, leafy browse and even mushrooms make up the bulk of the diet. During the fall, acorns are highly preferred in the few areas they are available.

Autumn ushers in the breeding season, which peaks during the first half of November. Males spend less time feeding and turn their attention toward establishing a position on the social ladder. A few weeks before the actual rut, adult male white-tails paw up pieces of ground about a square yard in size. They mark these scrapes with urine,

Right: White-tailed deer in favored hardwood habitat. Thomas W. Kitchin photo.

Left: The picture of innocence. White-tailed deer fawn approaches the photographer. Richard E. Kirchner photo   Above: A white-tailed doe waves flag of warning, not surrender. Richard E. Kirchner photo   Right: Once thought to be extinct, several hundred Columbian white-tailed deer enjoy sanctuary in Washington's Columbia White-tailed Deer National Wildlife Refuge. Thomas W. Kitchin photo.

and nearby trees often are scarred by their thrashing antlers, forming rubs. If another buck strays within such a breeding territory, he usually will be challenged. The gnashing of antlers can be heard hundreds of yards away, and sometimes other bucks are drawn to the fracas.

A doe passing by a scrape also may urinate in it, which informs males of her breeding condition. Whenever the male detects a "positive" scent in his scrape, he trails the doe. Along the way several males may join the procession, with the dominant individual taking the lead. After a doe is serviced, commonly by several bucks, they lose interest in her and begin searching for another.

Winter-time groups disperse as the fawning season arrives in late May and early June, following an approximate 200-day gestation period. One- or two-year-old does usually have just one offspring, while older females often produce twins if food supplies have been adequate. During the first few weeks, fawns remain tucked away in tall vegetation when the doe leaves for short feeding periods. Young ones begin nibbling on grasses and browse as early as two weeks of age. The importance of such food items gradually increases until the end of the fawn's fourth month, when they become its sole sustenance. Coyotes, bobcats and occasionally a black bear may capture a fawn. Cougars and wolves feed on adult deer, but these predators now seldom live where white-tails are abundant.

Experienced deer hunters know that white-tails prefer brushy hardwood cover along river and stream valleys. They also may be found at moderate elevations in immature forests with a shrubby understory and numerous meadow openings. Fires and logging help to improve some coniferous forest areas as white-tail habitat by opening up the dense forest cover, which allows food plants to thrive. Logging and agricultural clearing seem to have improved white-tail habitat in many areas, because this animal's range evidently has expanded over the last 100 years, sometimes at the expense of mule deer range. In the Northwest the

16

white-tail's range covers most of Montana and Idaho, with scattered populations in all but northwestern Washington and southwest and southeast Oregon.

White-tailed deer have slender bodies, long legs and a foot-long bushy tail, pure white on the underside. A large full-grown buck may weigh between 190 to 250 pounds while females weigh between 110 to 160 pounds. In adult males each antler is composed of one main beam from which a series of unbranched tines rises. Antlers grow from April to August and are shed between December and the latter part of February.

The Columbian white-tailed deer subspecies (*Odocoileus virginianus leucurus*) is considerably smaller than its relative. Coloration and thickness of its pelage varies according to season. Its sparse, reddish-brown summer coat is replaced by longer, thicker gray hair during the fall. Although now rare, the Columbian white-tailed deer was relatively common when Lewis and Clark made their way along the Columbia River from The Dalles to its mouth. Before the region was settled, this small deer ranged north from the Umpqua River near Roseburg, Oregon through the Willamette Valley to Puget Sound. Pioneers converted brushy, but fertile riverbottoms to agricultural fields, destroying the animal's essential habitat. By the 1930s the Columbian white-tailed deer was presumed to be extinct.

Fortunately, a few survived and reproduced, and by 1970 about 200 were hanging on. In 1972, to protect the remaining herd, the U.S. Fish and Wildlife Service created the 4,400-acre Columbian White-tailed Deer National Wildlife Refuge near Cathlamet, Washington on the Lower Columbia. Three years later the mammal received official federal designation as an endangered species, which meant even more intensive management. Today about 400 animals are scattered among the mainland refuge and Puget Island, and around Westport and the Tenasillahe Island refuge on the Oregon side. Another small herd lives along the Umpqua River just east of Roseburg, Oregon. Wildlife managers, through continued protection and possible transplants to other suitable habitat, hope to have this deer off the "endangered" list in the not-too-distant future.

Mule Deer and Black-Tailed Deer

Black-tailed deer, Audubon lithograph. Audubon, Quadrupeds. *Photo from Special Collections Division, University of Washington Libraries.*

Seldom is the exact origin of the common English name for a familiar animal known. With respect to the West's abundant mule deer, however, this is just the case. On May 10, 1805, Meriwether Lewis of the famed Corps of Discovery to the Northwest, recorded how the christening came about. He wrote, "The ear and tail of this anamal when compared with those of the common deer, so well comported with those of the mule when compared with the horse, that we have by way of distinction adapted the appellation of the mule deer, which I think is much more appropriate." This was part of the first detailed written description of the animal, and the name has been used since.

The explorer also noted that the French used the term, black-tailed deer, for mule deer, but he reserved that name for the smaller race of mule deer he encountered near the mouth of the Columbia River. He called it a "Black-tailed fallow deer," thus originating the common name for this previously undescribed subspecies. During the next 180 years, the common names for these animals have remained the same, while their scientific names have been changed time after time.

Mule deer (*Odocoileus hemionus*) live throughout the Northwest as far west as the Cascades. Black-tailed deer (*Odocoileus hemionus columbianus*) range from these mountains west to the beaches of the Pacific Ocean. Within the Cascades, hybrids of the two deer are relatively common.

Both mule and black-tailed deer have stocky bodies, long legs and large ears. The tail of the mule deer is thin and rope-like, with a black tip at the end of an otherwise white appendage. Its cousin has a considerably bushier, black tail. In both deer, the underside of the tail is white. Muleys inhabiting the Rocky Mountains are the largest of their kind with bucks averaging a little less than six feet long and slightly more than three feet high at the shoulder. Adult males typically weigh between 200 and 275 pounds when living on good range, while does seldom exceed 180 pounds. Most black-tailed deer weigh 50 to 60 pounds less than their relatives of the same sex. In summer, these ungulates are reddish-brown. They turn dark gray

with a blackish forehead and brisket at the onset of winter.

Only male deer normally have antlers. Adult mule and black-tailed deer both have evenly forked branching in their antlers, which tends to form four major points on each side. The number of points is subject to considerable variation. New antlers begin as buds in April or May, and an amazing growth rate of nearly a half-inch per day is maintained through spring and summer. Velvet is shed in August and September and most bucks lose their antlers in the latter part of January or early February, well after the breeding season.

Mountain foothills with relatively open coniferous forest are typical mule deer habitat. They also can be found around brushy coulees cutting through grasslands, in steep and rough terrain adjacent to shrublands and subalpine forests up to 8,000 feet in elevation. The wise hunter who is after the biggest bucks concentrates early season efforts in the highest mountains, at treeline or even above. Black-tailed deer favor the brushy edges created when farmland and meadows are interspersed with thick forests. Logged-over areas in early regrowth stages appear to be to their liking as well. These deer also take refuge in the isolated tree islands dotting Oregon's coastal dunes.

In the colder mountainous parts of the Northwest, depth of snow cover probably is the single most important weather parameter influencing mule deer behavior. During the warm season, mule deer are widely scattered over suitable terrain at higher elevations. Each individual establishes a home range of one to three square miles. The domains of males often lie on the periphery of those inhabited by females, but home ranges among individuals of the same sex may be largely overlapping.

When snow accumulates to depths exceeding six to 10 inches, deer start their migration to wintering grounds at lower elevations. The seasonal ranges may be separated by only a few miles or as many as a hundred. The survival of many mule deer that range over hundreds of square miles in summer may depend on critical winter range of only a few thousand acres. When human activities degrade such winter range, the effect can be disastrous for the whole herd.

Mule deer diets are highly variable. One study identified 788 *major* species of plants consumed by deer dwelling in the Rockies. Generally, grasses and forbs are consumed in spring and summer, while woody plants, including sagebrush and bitterbrush, make up the bulk of the diet during the snowy season. When winter browse is scarce, farmers' haystacks may be heavily used. In warmer coastal regions black-tailed deer prefer various browse species, such as trailing blackberry and red huckleberry. Thimbleberry is their mainstay during summer and early fall. Deer also may develop a taste for apples, and a good place to look for them in fall is an abandoned orchard. Rotting, fermented fruit can result in an inebriated animal, complete with belching, slobbering and disoriented staggering.

A mule or black-tailed deer that detects something unusual with its large ears or keen eyes, announces its alarm by whistling snorts. If it discovers real danger it speeds off in a characteristic bouncing gait, called stotting, trying to outrun the enemy rather than hide from it like a white-tailed deer. The animal bounces as high as two feet off the ground with all four feet bunched beneath the body. Each bound carries it 10 to 15 feet, and over short distances they reach speeds up to 35 miles per hour. When fleeing danger, these deer keep their tails down against the rump instead of waving them in the air like white-tails.

Both black-tailed and mule deer bucks may travel considerable distances at the onset of the rut. Antagonistic individuals prefer to employ various threat displays, such as snorting and thrashing their antlers in shrubs or trees rather than engaging in battle. Competing deer also use slit-like scent glands covered with tufts of long, coarse hair on the insides of their hind legs. During dominance disputes males sometimes bristle the hair, exposing the glands. The odor emitted from these metatarsal glands may intimidate other males.

The mule deer breeding season begins in early November, while black-tailed breeding starts in mid-October. By the end of December the rut is over for mule deer, while black-tailed bucks may chase does for a few more weeks. Fawns are born in June, after an average 200-day pregnancy. The gangly newborns arrive weighing six to nine pounds and remain spotted for nearly three months.

Fawns usually stay with their mother until one or two years old. However, in harsh environments, 75 percent of the fawns may not live to see their next summer. Black-tailed deer seldom live more than seven years. Male muleys rarely live past eight in the wild, but does may reach the ripe old age of 10 or 12.

Mountain lions, domestic dogs and coyotes are the major predators of these deer in the Northwest. In western Oregon research has shown that black-tailed deer are a staple of bobcats' diet. Occasionally, black bears may invite them to dinner. Most likely, automobiles take a greater toll among these animals than any predator, "deer crossing" signs notwithstanding.

Moose

Naturalists who study the journals of Lewis and Clark often are surprised to learn that not once did either of these explorers see a moose during their long journey to the Pacific and back. The closest they came to doing so was when one of their hunters wounded one in July 1806 at the headwaters of the Blackfoot River near Lincoln, Montana. The only other place moose were seen was in the Great Plains, a few miles up the Missouri from the mouth of the Milk River, where a hunting party spotted several. Considering the known range and habitat needs of the moose, this was an unlikely place to find the animal, even in pre-settlement days.

The present distribution of moose (*Alces alces*) in the Northwest is probably nearly the same as it was when trappers first scoured the region for beaver pelts. The Shiras moose, one of four recognized subspecies in North America, roams the mountainous parts of northern and central Idaho and western Montana. A few individuals inhabit the Okanogan Highlands and Selkirk Mountains of Washington, while some periodically cross the border from British Columbia to pay a short visit to the North Cascades. Although their historical range has not changed, their numbers have fluctuated drastically. As an example, around the turn of this century these ungulates were nearly extinct in Montana, and the hunting season was closed in 1897. By 1910 the herd had increased to about 300 animals. In following years, their numbers continued to grow rapidly and, when overgrazed ranges began to appear, the hunting season was reopened in 1945. Today several thousand can be found throughout Montana's western third, where hunters harvest about 500 annually.

Left: Bull moose antlers begin growing each March or April and attain full size by fall. Thomas W. Kitchin photo.

Above: Moose are most likely to be spotted foraging in shallow lakes and streams. Bruce Pitcher photo Right: With their exceptionally long, stilt-like front legs, moose must kneel to drink in shallow water. Richard E. Kirchner photo.

The moose is the largest representative of the deer family. Bulls average 10 feet in length and seven feet or more in height at the shoulders. They weigh between 800 and 1,200 pounds, while the majority of females tip the scales within the 600- to 800-pound range. Moose exhibit great weight fluctuations, however. An individual may increase its weight about 50 percent between early spring and autumn, followed by a gradual weight loss in winter.

Moose have an appearance only a mother could love. Their massive bodies with high, humped shoulders are supported by exceptionally long, stilt-like legs. Front legs are so long that animals must kneel to drink in shallow water. They have a long muzzle with a rounded pendulous nose, large ears and often a conspicuous bell of loose skin under the throat which sometimes freezes off in winter. The coat coloration varies from dark brown to almost black, but may become bleached toward the end of summer.

Only bull moose sport antlers, which have strongly palmated shovels. They begin their annual growth in March or April and attain full size around August or September, when the velvet is rubbed off against trees and shrubs. The exposed white antlers soon are stained brown from plant juices. Antlers of mature bulls between ages eight and 13 years are the largest. They may weigh 50 to 80 pounds and have a four- to five-foot spread.

The moose is the least sociable of all members in the North American deer family, spending most of the year alone or, in the case of adult females, in the company of calves. In early fall, bulls congregate in rutting groups to establish their social ranks. Males are aggressive at this time, and antagonistic encounters between individuals occur frequently. Occasionally an enraged bull may even send humans clambering up a tree for safety. During the actual rut, which extends from late September through early October, cows may join a male rutting group, consisting of up to 30 individuals, in search of a mate. Other females prefer more subtle means, such as a quavering call, to advertise their loneliness.

A cow in heat is tended constantly by bulls. If the pair encounters another male during their travels, the female waits nonchalantly for the outcome of the battle to determine which of the males will have the honor of her company. If the intruder is a cow, the roles are reversed, as the male remains a non-committed outsider. He displays his antlers and swollen neck and shoulders in front of the two females who are engaged in a dominance show-off. The winner joins the bull for the remainder of her heat.

During May or early June one or two calves arrive after an eight-month gestation period. The youngsters are reddish-brown at birth, but without spots. They are capable of swimming and can outrun a man within a few days. Calves have a well developed instinct to trail their mother, and they commonly stay with her until the birth of the next year's offspring. Moose calves increase their weight faster than any other ungulate, growing about 2-1/2 pounds each day during the first five months of their lives. From a 25-pound weakling at birth, there develops a formidable 375-pound youngster within five months!

Moose usually establish two home ranges; a larger one for summer, when moving about is easier, and a more limited one for the snowy months. The animals commonly spend the snow-free months in the higher country, and retreat to the lowland coniferous forests, which are interspersed with streams and plentiful willow thickets, for winter.

Some moose populations evidently use traditional migration routes between their seasonal ranges, which even may include swimming across lakes or rivers. The small town of Whitefish, in northwestern Montana, was built in the middle of such a route. The local moose herd winters on the southwestern side of town, but spends its summers to the northeast. The shortest route connecting the two ranges runs through town. Complaints about "giant ugly horses" running loose within the city limits are fairly frequent in spring and fall. Such incidents usually result in wildlife officials escorting the determined individuals to safer country along the river outside of town. The distances between such seasonal ranges are commonly less than 10 miles, although moose are known to spend winters in eastern Idaho, and each spring traverse 50 miles of rugged terrain to reach their familiar summer haunts in Montana's Centennial Valley or around Hebgen Lake. In other locations, moose herds may be relatively sedentary, remaining in the same area year-round.

The limit for preferred habitat of the Shiras moose closely follows the southern fringe of the northern coniferous forest. They prefer mixed hardwood-conifer forests with boggy areas and brushy stream bottoms, as well as mountain meadows and overgrown clear cuts. During summer their diet includes a wide variety of plants, ranging from mosses to leaves and twig tips. Browse, such as willow, is definitely a staple and almost never constitutes less than 50 percent of their forage. Various emergent and submerged aquatic plants are relished as well. In winter, more than 99 percent of the foods eaten are woody species, among which willow is most important in the Northwest.

Moose living in the Northwest have few natural enemies. They can, in most cases, easily outrun their adversaries, or convince them to leave with a few well-placed strikes of their front hooves. Traditionally, wolves were the most important natural predators of these ungulates, while a few occasionally may have succumbed to black bears and grizzlies. Loss of wintering habitat, caused by the invasion of homes and recreational developments into mountain stream valleys, is the most serious threat to moose survival today.

Woodland Caribou

Almost half the entire Selkirk caribou herd grazes near the Idaho-British Columbia border. J. Bart Rayniak photo.

Years ago, Gene Autry made a fortune singing his hit song, "Rudolph the Red-nosed Reindeer." Maybe the West needs another singing cowboy to help raise money for the preservation of Rudolph's cousin, the mountain caribou.

The mountain caribou (*Rangifer tarandus*) is considered by some to be a distinct variety of the woodland caribou, and both are close relatives of the European domestic reindeer. An isolated population of mountain caribou inhabits several hundred square miles of the mature boreal forests and alpine tundra of the Selkirk Mountains in northeastern Washington, northern Idaho and southern British Columbia. Some also may visit extreme northwestern Montana.

Before a large portion of this area's boggy climax forests was eliminated by homesteading, agricultural activities, logging and burning, the range of the mountain caribou extended as far south as Missoula, Montana and into some of the drainages of central Idaho. The herd now is estimated to include between 25 and 30 individuals, only a fraction of the 100 to 200 animals found in the region as recently as the 1950s. There may be only seven or eight breeding females, and the herd is barely holding its own.

When British Columbia completed the construction of Highway 3 through the herd's range in 1963, several animals became highway fatalities, probably because of their unfamiliarity with this danger and the appeal of highway salt. Hunters evidently have had trouble distinguishing female caribou from deer or elk, and at least four caribou cows have been shot in the last five years. Because caribou numbers continued to dwindle, they were granted endangered species status in the United States in January 1983, despite opposition by the U.S. Forest Service. It now is a federal offense to kill one of the animals, with a maximum penalty of up to one year in prison and a $20,000 fine. Canadian officials also passed laws to protect the herd, thus enhancing chances for survival of the last caribou in the contiguous United States.

The woodland caribou subspecies is generally darker in color and larger than its Arctic relatives in North America, the barren ground caribou. The mountain caribou is larger than a big mule deer, mature bulls weighing from 350 to 500 pounds. During the winter the pelage is grayish-brown on the back, with the rump and undersides a grayish-white. In the cold season bulls also have a whitish collar and a thick, white mane. In summer animals are a uniform chocolate brown.

The caribou is the only member of the deer family in which both sexes have antlers. Those of older males can be impressive, but those of females resemble antlers of a spike deer. Antlers of mature bulls are asymmetrical with only one brow tine well developed and extending vertically over the face. The other brow tine is "abortive." Bulls normally shed their antlers soon after the October-November rut, although some exhausted individuals may drop them during the breeding season. Females and juvenile males retain their antlers throughout the winter, dropping them in the spring.

The hooves of mountain caribou are well adapted for the boggy meadows they frequent in summer and the crusted snow they prefer in winter. During the warm months hooves, which are as wide or wider than long, have spongy pads to carry the animals over soggy ground. As winter approaches, the soft pads become smaller and harder, leaving the rim of the hoof exposed. The sharp edge gives the caribou excellent traction on crusted snow.

Caribou in the Arctic migrate hundreds of miles each fall and spring between their seasonal

23

Members of North Idaho's Selkirk caribou herd are close relatives of the European domestic reindeer. J. Bart Rayniak photo.

ranges. In the Selkirk Mountains much shorter migrations are affected by changes in snow cover. The animals spend their summer and fall months foraging in high mountain meadows (over 5,000 to 6,000 feet) and adjacent open spruce and subalpine fir forests. They feed largely on sedges, grasses and forbs, although twigs, leaves of woody plants and mushrooms also are consumed. When the soft snow becomes so deep that it hinders caribou activity, they move to lower valleys, preferring wet areas within mixed hemlock and cedar forests. After the snow becomes crusted in the high country, the animals return to the wind-swept ridges of their summer range and remain there until spring.

During winter months mountain caribou subsist on lichens growing from tree limbs. These plants do not grow well unless host trees are 100 to 200 years old, and even then their growth rate is only a few tenths of an inch per year. Since caribou consume about 11 pounds of lichens each day, they must roam widely, incessantly in search of food. Wildlife biologists estimate that each individual requires about 150 acres of old-growth forest to meet the energy demands of one winter.

When the warming temperatures of spring render snow too soft for animals to walk on, they once again retreat to the lowlands, where they remain until snow melts from the alpine meadows in May or June. Thus, mountain caribou commonly spend their entire lives above 4,500 feet, and are one of

the few animals to endure the long, harsh winters of the alpine tundra.

Female caribou have a well synchronized estrous cycle, which means that 90 percent of all calves are born within a 10-day period in the latter part of May or early June, after a seven-month gestation period. Within a few hours of birth, the newborn is capable of following its mother. Although the calf begins to nibble on green vegetation at two days of age, it nurses until November. During its initial five months the youngster grows rapidly and increases its weight from 12 to 100 pounds, an amazing 800 percent! A mother usually remains with her offspring for a year, forcing it to leave just before the arrival of her next calf.

As the breeding season draws near in September, mature bulls, which are mostly solitary outside of rut, establish a dominance hierarchy. The largest males seldom engage in fights. Instead, they merely reinforce their social rank by showing off their formidable antlers and large bodies. Younger bulls of roughly equal size, however, often battle each other to determine their positions. During the October-November rut the highest-ranking bulls follow the cows in heat and barely take time to eat.

In the Selkirk Mountains natural predation on mountain caribou probably is light. However, rare grizzlies, wolves, wolverines and lynxes all are potential predators. Major threats to the Northwest's mountain caribou are continued destruction of habitat and loss of individuals due to accidents. Their precariously low numbers may make them susceptible to inbreeding, which can result in genetic deformities. One way to prevent inbreeding problems is to bring in outside animals. According to Mike Scott, wildlife biologist with the Idaho Department of Fish and Game, this option was being considered in 1985. Animals might be transplanted from the Revelstoke, British Columbia area to a site west of Bonners Ferry, Idaho. Some local citizens object to this, feeling that preservation of an endangered species may take precedence over local economic interests, particularly commercial logging. Even if a transplant does take place and is successful, the Selkirk caribou herd will be far from secure in the foreseeable future.

Pronghorn Antelope

Purists sometimes declare that if a North American hunter wants to go antelope hunting, he or she should book a flight to East Africa or Southeast Asia. The point being that the North American pronghorn is not closely related to the true Old World antelopes, such as gazelles. The pronghorn is so unique that most taxonomists place this animal in its own separate mammalian family. Use of the term, antelope, for this animal seems to have orginated with Lewis and Clark, who provided the first accurate written description. Clark wrote, ". . . he is more like the Antilope or Gazella of Africa than any other species of Goat."

Above: Pronghorn antelope, resident of the Northwest's open prairies and sagebrush plains. Alan Carey photo Right: Pronghorn antelope, Audubon lithograph. Audubon, Quadrupeds. Photo from Special Collections Division, University of Washington Libraries.

On Stone by Wm E. Hitchcock

The pronghorn (*Antilocapra americana*) is the only surviving member of a group of animals that has been evolving in North America for 20 million years. There is some question whether pronghorn have true hair horns, even though the horny material may share a common origin with hair. The horns differ from those of all other ungulates in two major ways. They are shed and regrown annually, and they are the only horns with a fork, the prong. Adult males typically have 10- to 16-inch-long horns, curved at the tips. About 70 percent of does also carry horns, but they are considerably smaller, averaging just 1-1/2 inches long. Males usually shed the horny sheaths in November or December to make way for next year's set. The new horns grow for 9 to 10 months, reaching their maximum development in August or September.

Even without their unusual horns, pronghorns are easy to distinguish from all other North American big game species. Their deer-like bodies are reddish-tan on the back and white underneath, with a large white rump patch. They have extremely large and protruding eyes, which provide the animal with an extraordinarily wide field of vision. Black cheek patches adorn males but are absent in females. Pronghorns usually are about 50 inches long and up to 40 inches tall at the shoulder. Bucks average 125 pounds and does 110.

During winter pronghorns form herds containing dozens to hundreds of animals of both sexes and all age classes. As spring arrives these congregations break up into smaller bands of does, bachelor groups of bucks and solitary older males. Yearling or older does that bred the previous fall commonly deliver a set of twins in May or June. Newborn fawns are uniformly grayish-brown and weigh between six and nine pounds. They are able to walk within 30 minutes of birth and, within a couple of days, are capable of outrunning a man. Young normally stay hidden in the vegetation while the mother grazes close by. After fawns turn three weeks old they begin following the females during daily foraging. At this time several does and their youngsters join together in nursery herds.

Increased protection against predators is a prime reason pronghorns form groups. Whenever one individual detects danger, it flares its white

Pronghorn doe nurses one of her twins. Alan Carey photo.

rump patch, signaling others to flee. Pronghorns are well adapted for outrunning pursuers. Their oversized windpipes carry copious amounts of oxygen to unusually large lungs, which sustain sprints of 45 to 50 miles per hour. Such speed, together with a keen visual "advance warning system," makes adults difficult prey for any natural predator. Fawns may be taken by coyotes and bobcats. Adults weakened by severe winter weather also may fall prey to these predators, as well as to free-roaming dogs.

While the pronghorn's speed frustrates most predators, it is no match for a tight fence. Contrary to widespread belief, pronghorns can jump fences, but they are reluctant to do so. They would rather crawl under or through a fence. Those made of woven wire or four to five tight strands of barbed wire may corral pronghorns, allowing capture.

The pronghorn breeding season commences in mid-September and extends through October. During the rut older males defend territories in areas having the best food supplies. They warn any intruding buck with loud snorts and wheezing coughs. If this does not scare off the opponent, a fight may ensue. Contenders slowly approach one another until horns meet, triggering vigorous twisting and shoving. Eventually the weaker individual retreats. Although fights may be bloody, fatalities are rare. Bucks reigning over the choicest territories attract the largest number of female visitors and do most of the breeding.

A pronghorn's day is divided among feeding, resting, ruminating and trips to water holes in dry areas. Browse species, like sagebrush and bitterbrush are the most important foods year-round and become critical during winter. Succulent forbs are consumed in spring and summer. Pronghorns may actually improve some rangelands by eating vegetation poisonous or unpalatable to livestock. They seem to relish loco weed, lupine, thistles and even prickly pear cactus. Grasses appear to be the least-used food items, but may be eaten in early spring when the young and tender shoots are of high nutritional value.

Nursery and bachelor herds normally forage within two- to five-square-mile home ranges during summer, while solitary bucks roam over a territory one-third or less that size. In winter, home ranges of the herds are smaller. Pronghorns also may have separate seasonal ranges. For example, in Oregon some animals spend summers on the high sagebrush plateaus along the eastern slope of Hart Mountain within the 350-square-mile Hart Mountain National Antelope Refuge. At the onset

of winter they migrate 30 miles south to the Sheldon National Wildlife Refuge across the border in Nevada.

In the early 19th century pronghorns ranked second only to bison in numbers, with an estimated 35 million throughout the West. The herds soon were decimated by conversion of rangeland to cropland, professional hunters who sold the meat and ranchers who erroneously believed the creatures were seriously competing for forage. By the 1920s, it was estimated that a mere 15,000 to 20,000 individuals remained, with 2,000 to 3,000 each in Idaho, Montana and Oregon. Today, thanks to transplant programs and careful management, the pronghorn roams again in the sagebrush prairies in herds totaling over half a million. Their present range in the Northwest covers most of the suitable habitat in eastern Oregon, southern Idaho and southwestern Montana.

Apparently pronghorns were never common in Washington, although the state appears to have

some suitable habitat. By the time white men settled the state, the species was virtually extinct. Between 1938 and 1940, 138 pronghorns were obtained from Oregon's Hart Mountain National Antelope Refuge and Nevada's Sheldon National Wildlife Refuge. Another 21 animals came from the Burns, Oregon area in 1968. Releases were made in Kittitas, Grant and Adams counties in central Washington. Today, possibly 30 live in the vicinity of the Department of Game Colockum Habitat Management Area north of Ellensburg, and fewer than 10 survive on the U.S. Military Reservation Yakima Firing Range northeast of Yakima. Small bands possibly exist in the Beezley Hills west of Ephrata, and in the Ritzville area. Biologists with the Washington Department of Game in Yakima note that the animals seem to be barely hanging on, but the reasons for lack of success are unknown. Evidently, pronghorns still have not changed their minds about making a home in the Evergreen State.

Below left: Common in sections of southern Idaho and eastern Oregon, pronghorns are scarce in the dry plains of eastern Washington. Richard E. Kirchner photo Below: Seemingly useless sagebrush provides important year-round sustenance for pronghorns. Richard E. Kirchner photo.

Bison

White trappers and fur traders of the early 19th century arrived just a few years too *late* to witness the extermination of the bison in much of the Northwest. As early as 1805, Meriwether Lewis noted that the Jefferson River Valley near Three Forks, Montana contained nothing but bones of bison. A year later in the same locality, Sacajawea told William Clark that bison had been numerous in the area years before. Likewise, trappers visiting eastern Washington, southern Idaho, and southeastern Oregon during 1810 to 1830 obtained only indirect evidence that bison once had been fairly common. Ross Cox, who worked for the North West Company during 1813 to 1817, related that Indians informed him bison once had been numerous in eastern Washington, and their remains could still be found. Hudson's Bay Company employee Peter Skene Ogden, during an 1826 visit to Harney and Malheur lakes in southeastern Oregon, found many bison skulls but no living animals.

Archeological data compiled by Gerald F. Schroedl of Washington State University indicated that, at least in eastern Washington, Indians made the greatest use of bison (*Bison bison*) during 500 B.C. to 500 A.D., and that they were common at least until 1500. In these early times, hunters were horseless. Killing the huge mammals was facilitated by driving them over cliffs (buffalo jumps), herding them into constructed enclosures (pounds), stalking by individuals or surrounding them with groups of hunters. Between 1500 and 1800, aboriginal use diminished considerably in eastern Washington. Zoologist Gene M. Christman noted that Indians of the Columbian Plateau first obtained horses by about 1720, and speculated that this new mobility enabled the natives to exterminate bison over a large area.

Bison inhabiting eastern Washington, southeastern Oregon and isolated mountain valleys in northern Idaho and western Montana evidently were overflow from the main population centers of the plains, and were occupying marginal habitat.

Left: Bison, Audubon lithograph. Audubon, Quadrupeds. *Photo from Special Collections Division, University of Washington Libraries Below: In 1985 the Montana legislature declared the bison a game animal and ordered a hunting season as a means of controlling bison wandering out of Yellowstone National Park. Bruce Pitcher photo.*

As these were wiped out equestrian natives from the Nez Perce, Flathead, Coeur d'Alene and Spokane tribes began making forays to the plains to secure bison. At about this time white trappers arrived in the Northwest. By then a few isolated herds remained in Montana's larger southwestern valleys. The Snake River Plains and valleys of the Salmon River headwaters in Idaho probably had several tens of thousands of animals, based on early 19th century observations of mountain men, including Alexander Ross and Peter Skene Ogden.

Some historians and zoologists believe that the bison of the mountainous western region were slightly different from the millions on the plains to the east. The mountain, or woodland, subspecies was supposedly larger and darker-colored than its plains cousin. However, geographic separation of these races does not seem to be supported by the observations of early explorers. Both David Thompson and Colonel J. J. Abert noted that plains bison ventured into the mountains.

Regardless of their subspecific status, the bison of Idaho and western Montana apparently were some of the first Far West herds to be eliminated by white men. Abert and his colleagues reported bison rapidly disappearing on the Snake River Plains about 1834-35, and that the huge beasts were gone by 1843-44. Dr. George Suckley, naturalist for the Northern Railroad Survey, reported bison were exterminated in the Bitterroot Valley by 1853.

Left: Most Northwest bison today are behind barbed wire, a stark contrast with their free-roaming ancestors. John A. Alwin photo.

By 1890 the only significant wild bison herd left in the U.S. was the few hundred surviving within the confines of Yellowstone National Park, and even they were under intense poaching pressure. U.S. Army troops were stationed in Yellowstone to help protect the herd, and Congress passed a law in 1894 that levied a $1,000 fine for anyone killing a wild bison. In 1905 the American Bison Society was formed for the preservation of the animals and soon obtained about 18,000 acres of somewhat marginal bison habitat from the Flathead Indians. In 1909, 41 bison were stocked on what became the National Bison Range near Moiese, Montana. Today there are 300 to 400 bison at Moiese. The annual first-week-of-October roundup, necessary to thin the herd to keep it in balance with carrying capacity, has been a major tourist attraction for decades.

People who are quick to condemn our forefathers for nearly exterminating the bison seem unaware that human economic attitudes have changed little since then. Bison have a natural wandering instinct, and every spring a few of Yellowstone National Park's estimated 2,000 animals attempt to leave the overgrazed park. In February and March of 1985, Montana Department of Fish, Wildlife and Parks officials were required to kill 88 bison headed north to better (private) pastures in the Paradise Valley south of Livingston. The Butte Skyline Sportsmen's Club suggested a special hunting season was a better way to solve the problem. Regardless of the methods used, unsympathetic ranchers wanted the bison removed immediately to prevent fence damage and forage loss. Also, it has been presumed, but not documented, that these bison would infect cattle with brucellosis, which can cause calf abortions.

The very nature of the bison as a prolific wild mammal assured its doom in the face of advancing civilization. If the hide hunters hadn't killed them, someone else would have. The American buffalo is the largest terrestrial mammal native to North America. Big fence-busting bulls measure 10 to 12 feet in length, 64 to 72 inches in height and weigh between 1,000 and 2,200 pounds, while cows are about 25 to 30 percent smaller. Natural control of bison numbers is no longer possible. The only important predators of bison were wolves, which generally were only able to bring down the large animals through cooperative pack effort. Grizzly bears were thought to kill some, especially in spring after the "shaggies" were weakened by a long, hard winter.

Bison spend most of the day grazing, much like domestic cattle. Various grasses, forbs and sedges make up the bulk of the diet. Only after preferred food items become scarce will the animals feed on woody browse. When bison numbered in the millions, herds probably ranged over several hundred square miles, moving from one seasonal range to another, which reduced problems of overgrazing.

Ranchers willing to put up with their independent attitude have found they make valuable meat animals. Jerry Nyhart and his wife, Wink, produce 25 to 30 bison per year for slaughter on their ranch 10 miles southwest of Twin Bridges, Montana. They affirm that the beasts can be difficult to fence or herd, but this is offset by the facts that they are relatively disease-free, and that calving is well taken care of by the cows. Other ranchers have crossed bison with cattle to produce faster-growing "beefaloes," or "cattaloes." The plains are no longer darkened with vast bison herds, but there are at least 80,000 animals in North America, many kept by livestock producers.

29

Mountain Goat

Heavy snow on a steep mountainside can be the mountain goat's Grim Reaper. Nevertheless, when an avalanche freight-trained down Running Rabbit Mountain on Glacier Park's southern boundary in early 1979, it was the proverbial blessing in disguise for local goats. A very ordinary two-lane steel bridge on U.S. Highway 2 had the misfortune of being in nature's way, and ended up several hundred feet below on the bank of the Middle Fork of the Flathead River.

When federal and state highway officials decided to rebuild the span and adjacent roadway, all agreed it was time to consider goat needs. For years, the busy highway had cut across a migratory path the animals followed to reach a natural salt lick on the banks of the river. Mountain goats from up to 15 miles inside the park had to dodge cars and curious photographers just before they reached their saline goal. When the new concrete bridge and road were finished in 1981, special provisions were made to include a goat underpass. Fence leads even were constructed on the side of the mountain to help guide the goats to their new walkway. Today just a few miles southeast of Essex, Montana, tourists can park in a small "Goat Lick Overlook" lot and watch dozens of the unmolested animals go about their business.

When not concentrated at scarce salt licks, mountain goats (*Oreamnos americana*) are less gregarious than most ungulates. Bands of billies, or nannies and immature animals, seldom consist of more than five individuals. After kids are born, several females temporarily join together, creating fairly large nursery bands which offer increased protection against predators.

In late October the billies begin to follow the females, who repeatedly reject their advances. During the rut, fights between males are uncommon. Various threat displays, like bluff attacks, are employed to suppress aggressive behavior. Males also paw the ground with their front legs, sending soil and snow flying and creating a rutting pit. If a battle takes place, the opponents circle one another, stabbing at each other's belly, flanks and rump, instead of clashing head-to-head like most ungulates. Toward the end of November the females are ready to breed and the rut is over by early December.

After an approximate six-month gestation period the usually single kid arrives. Nannies retreat to inaccessible cliffs to give birth and within a few hours the youngster is capable of climbing around the steep slopes. The mother weans her offspring 8 to 10 weeks later, but it accompanies her for nearly a full year, until the next kid is born. Without mishaps, a wild mountain goat may live 13 years.

Mountain goats spend their summers grazing on succulent plants in alpine and subalpine meadows adjacent to the steep cliffs necessary for escape. If the area contains ridges and slopes swept clean of snow by blowing winds, they will remain above timberline year-round. When such terrain is absent, the creatures descend to lower elevations for the snowy season. The most important food items in the summer are various grasses and sedges, although forbs such as lupines and mountain bluebells also are consumed. Winter diet varies according to availability. In areas where grasses can be found, they continue to be important. When they are lacking, goats become browsers and rely heavily on woody materials. Throughout all seasons mosses and lichens are consumed in small amounts.

Mountain goats have a vertically-compressed body with short legs. The spongy pads of the hooves are slightly convex, and they extend outside the hard outer shell of each toe to give the animals better traction on cliffs. Goats have a prominent hump at the shoulders. Males measure around 42 inches high and 70 inches in overall length; nannies average 160 pounds, while males average 180 pounds. Adult goats have a long, shaggy yellowish-white coat that covers the entire animal except the lower legs and face. The narrow head with a black muzzle is adorned by a pair of large dark eyes, ears that are big and pointed, and a prominent beard. Both sexes have conical, dagger-like black horns.

Natural enemies of mountain goats are few, and predation seldom is significant. The main threat to small kids is diving golden eagles, which may cause them to fall from cliffs. The birds then feast on the carcass. Occasionally a coyote gets a youngster, while mountain lions and wolverines can prey on adults. Some goats are swept away by avalanches and rockslides, and many are injured from falls in their precarious, nearly vertical, habitat.

Because of its rugged home, the mountain goat is a challenging game species, but limited populations mean only 700 to 800 fortunate Northwest hunters who win in state drawings are allowed to pursue this high mountain quarry. Lewis and Clark were unable to see one up close but, based on skins supplied by the Indians, they called it a sheep. Even today, the genus name for the animal is derived from the Greek roots *ore* (a mountain) and *amno* (a lamb).

Mountain goat nanny and kid perched high atop their Glacier National Park world. Alan Carey photo.

This animal originally inhabited the highest peaks in the main mountain chains of the Northern Rockies and Cascades. Its remote habitat largely has been unaffected by civilization and, in fact, this is one of the very few species whose range has been greatly increased by man's intentional efforts. All four Northwestern states have transplanted mountain goats.

In the Olympic Mountains of western Washington, the dozen transplanted goats from Alaska and British Columbia proved too successful. For several reasons, including continental glaciation, the Olympic Peninsula has existed in relative biological isolation for thousands of years. Isolation meant that mountain goats, as well as the pika, lynx, wolverine, grizzly and mountain sheep did not naturally inhabit the peninsula. Even though not native, the goats thrived and now total nearly 1,200. Such large numbers of mountain goats are severely impacting the park's fragile alpine and subalpine areas. Large and numerous wallows used for dust baths are the most visible and significant consequence of the goats' presence in the high country. An estimated 45 tons of soil have been displaced from just one wallow on the park's Klahhane Ridge. In the mid-1980s the Park Service was experimenting with various live-capture techniques, and looking for other areas in the market for mountain goats.

Above left: Spotting for mountain goats is a popular recreational activity in the Northwest. This goat range along U.S. Highway 12 in northern Idaho is one of many hot spots. M. Douglas Scott photo   Above: Strait of Juan de Fuca sunset provides a backdrop for this Olympic Mountains resident. Thomas W. Kitchin photo   Above right: Spongy pads on hooves help give mountain goats the traction necessary for moving about in their rocky, near-vertical home. Richard E. Kirchner photo   Far right: Mountain goat partakes of high country vegetation. Richard E. Kirchner photo   Right: Within a few hours of birth, a mountain goat kid is capable of climbing steep slopes. Richard E. Kirchner photo.

Bighorn Sheep

About the time snow begins to accumulate on the Northwest's high mountain ridges in early November, a thundering crash may reverberate throughout the steep, rocky slopes and lower valleys. This storm of massive colliding horns marks the onset of the bighorn sheep (*Ovis canadensis*) breeding season. Adult males are in the process of establishing a dominance hierarchy that determines which individuals will take care of most of the breeding.

Bighorn rams can boast curled horns bigger in proportion to their body size than those of any other ruminant. An older animal's may measure 45 inches in length, 15 inches in circumference at the base and account for 8 to 12 percent of total body weight. It is no wonder that many hunters consider the bighorn ram to be one of the most impressive trophies on the North American continent. Regional outfitters may receive up to $8,000 for a guided sheep hunt, successful or not.

Although sheep horns grow throughout the year, the process intensifies in summer. Distinct annual rings, which can be used to estimate a ram's age, are a result of this growth spurt. Horn size alone is often enough to deter a younger ram from challenging an older individual. When two sheep of roughly equal size meet, however, a battle is likely to ensue. At speeds as high as 30 miles per hour, the two jousters charge at each other, sending dirt flying as they collide head-on. Such competition is sometimes settled after the initial bout, but more commonly the animals will clash several times before the matter of superiority is clear. One observed duel went on for more than 25 skull-slamming hours! Eventually one ram surrenders, either because of exhaustion or painful injuries.

The most successful of the battling rams moves from one band of ewes to another in search of receptive females. They breed as many as their stamina will allow before the end of the rut in late December. After a six-month gestation period the single young arrives. Lambing commences in late April and lasts through June. The lambs are born well developed and coated with soft, woolly hair. Within a day they are capable of following the

ewe, climbing the precipitous cliffs almost as well as she. By the end of the second week they nibble tender shoots of grass, though they continue to suckle for another 4-1/2 months.

The price of success is high in bighorn sheep society. Most of the dominant rams die before their 13th birthday, while the less successful may enjoy life's pleasures for 17 to 19 years. The lifespan of the peaceful ewes can be as long as 24 years.

Mountain sheep feed on bunch grasses and forbs throughout the summer. In winter sustenance is derived from woody browse, such as sagebrush, rabbitbrush and willow. If the snow is not too deep the animals often will paw through it in search of grass. Following the lambing season, sheep move up-country to spend summers in verdant alpine meadows. When migrating to the lower and drier winter ranges, older sheep lead the young along routes believed to have been used by the herd for generations. The distances traveled

Stocky bighorn sheep only slightly resemble their domestic relatives. Bruce Pitcher photo.

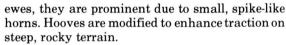

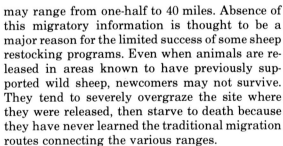

Above: Bighorn sheep (Argali) lithograph drawn by T. Doughty, 1830. Partly based on specimens collected by Captain Meriwether Lewis in western Montana. Vol. 1, plate 17 of J. Doughty and T. Doughty, eds., The Cabinet of Natural History, American Rural Sports with Illustrations *(3 vols.; Philadelphia: J. & T. Doughty, 1830-1833). Photo from Library, The Academy of Natural Sciences of Philadelphia   Right: The thunder of colliding horns signals the onset of the bighorn breeding season. Alan Carey photo.*

may range from one-half to 40 miles. Absence of this migratory information is thought to be a major reason for the limited success of some sheep restocking programs. Even when animals are released in areas known to have previously supported wild sheep, newcomers may not survive. They tend to severely overgraze the site where they were released, then starve to death because they have never learned the traditional migration routes connecting the various ranges.

Bighorn sheep rely on their keen eyesight to detect anything unusual in the surrounding open areas. If alarmed, the feeding animals quickly take to nearby inaccessible cliffs. Mortality due to predation is light, the major threats being wolves and mountain lions. In rare circumstances coyotes take a few individuals. A hungry golden eagle every now and then may swoop at a lamb, but they seldom achieve their goal, because the little one

quickly seeks shelter under the belly of its mother.

A more serious threat to the well-being of mountain sheep is competition for forage with livestock or even mule deer and elk. Most commonly the conflicts arise between wild and domestic sheep, but in some areas in Idaho, cattle have depleted mountain sheep forage. Bighorns also are susceptible to several serious diseases, such as anthrax, pink eye, scabies mites and pneumonia.

Bighorns are stocky mammals with only a slight resemblance to the domestic sheep. Their overall body coloration is grayish-brown, while the muzzle, rump patch and undersides are off-white. Rams weigh about 300 pounds, 100 to 200 pounds more than the females. The average males measure five feet in length, while ewes are about half a foot shorter. The massive, spiraled brown horns of the rams make their erect ears almost unnoticeable, but on immature animals, as well as on

ewes, they are prominent due to small, spike-like horns. Hooves are modified to enhance traction on steep, rocky terrain.

Bighorns formerly ranged throughout most of the Northwest's higher mountains, as far west as, but evidently not including, the Cascades. They also did well on isolated rocky buttes, and cliffs along major rivers, far out in the semi-arid plains. Due to disease, livestock competition, and possibly over-hunting, sheep numbers reached a low ebb in the Northwest during the 1930s, even to the point of extermination in Washington.

Federal firearms and ammunition tax revenues enabled Montana to start transplanting the Rocky Mountain race of bighorn sheep to vacant ranges during the 1940s. In the '50s and '60s, British Columbia supplied the California bighorn sheep subspecies for restocking in central Washington, eastern Oregon and southern Idaho. Restocking of

both races continued into the '70s and '80s, and now most suitable bighorn habitats have at least a few sheep. All four states now offer limited bighorn sheep hunting by special permit, evidence that careful wildlife management in some cases can be successful in restoring indigenous species. Usually determined by a draw and highly prized by Northwest hunters, permits in recent years have allowed annual hunts of 400 to 500 animals, the majority from western Montana. In Washington, where commonly less than one percent of applicants are successful in the draw, odds are overwhelmingly against any hunter winning an opportunity to bag this largest and best-known wild sheep of the North American continent.

Above: Bighorn rams sport exceptionally large horns in proportion to their body size. Richard E. Kirchner photo Above right: Bighorn lamb. Richard E. Kirchner photo Right: California variety of bighorn sheep in northern Washington's Okanogan country, late 1950s transplants from British Columbia. This spring shot catches the animals at their least becoming, shedding winter hair. Thomas W. Kitchin photo.

Carnivores

Young long-tailed weasels with mouse. Alan Carey photo.

Coyote

With the sometime name of "watermelon wolf" you might guess the coyote is a highly adaptable carnivore. On the outskirts of Boise, Idaho, a family carefully cultivated a small garden. One section was devoted to watermelons, everyone's favorite. After much watering and weeding, it was decided that the melons were ripe enough to begin harvesting the following day. As if aware of the decision, the local coyotes raided the patch that night. To make matters worse, instead of finishing off just a few of the fruit, they took several juicy bites out of every one, rendering them nearly useless to the unhappy gardeners.

This clever, doglike mammal played a prominent role in Native American culture. Many tribes gave it credit for changing the ancient world and its prehistoric creatures to those inhabiting Mother Earth today. The coyote also was held responsible by some aborigines for the creation of the Indians.

European immigrants moving to the Northwest viewed the animals in a different light. They regarded the "brush wolf," also called the "little wolf," as an undesirable competitor that killed both valuable livestock and desirable game species, such as deer. Coyotes (*Canis latrans*) were subjected to any eradication methods imaginable, including use of the poisons, strychnine and 1080. Despite this relentless persecution, the resilient coyote has thrived, earning grudging admiration from some ranchers.

Before white settlers logged the dense climax forests west of the Cascades and cleared other areas within that region to create farm fields, the coyote was uncommon in both western Oregon and Washington. As evidence of this fact, Indian tribes living west of the Cascades had almost no legends about the coyote, but natives to the east knew the animal well. The original coyote range included the more open, arid interiors of the northwestern states, along with the foothills, subalpine and alpine terrain of the Northern Rockies. Opening of mature forests enabled this species to ex-

Blue-eyed coyote pups. Alan Carey photo.

tend its range clear to the Pacific Coast. Today coyotes can be found throughout the Northwest, even as suburban residents outside some large cities.

Coyotes are slender animals resembling the wolf, though considerably smaller. Their bodies measure about four feet from the nose to the end of the usually black-tipped tail. Average height at the shoulders is two feet, and most weigh between 20 and 30 pounds. Their thick, long and glossy winter pelage varies from reddish-gray to grizzled-gray with the undersides and legs a dusky white.

Social organization among coyotes varies greatly from loners to gregarious groups inhabiting a specific area for a long time. This variability is believed to reflect the nature of food availabil-

ity. Small packs most commonly are formed by a mated pair with whom some of the offspring remain. This occurs at the onset of winter in areas where carrion of large ungulates, such as elk and deer, make up the bulk of the diet. A coyote pack does not collectively hunt prey as do wolves. Instead the additional members defend carrion from other scavengers. During summer months when small mammals are the main food items, hunting easily is conducted by an individual.

Coyotes are highly opportunistic carnivores with healthy appetites. Common food items include mice, ground squirrels, pocket gophers, rabbits, birds and their eggs, frogs, fish, insects, blackberries and even juniper berries during winter. Although it seems unlikely that the brush wolf

Right: The clever, doglike coyote was an important character in many Northwest Indian legends. Alan Carey photo Bottom: Wily and wary coyote investigates a suspicious sound. Bruce Pitcher photo Below: A winter's catch in northeastern Oregon, early 20th century. Joseph Museum, Oregon Historical Society photo, neg. no. 26832.

would bypass an opportunity such as an unprotected deer or antelope fawn, studies have indicated that the losses they cause among wild game are not severe enough to justify spending large amounts of money on control programs.

Coyote predation of sheep, however, can be considerable. Dr. Bart O'Gara and his colleagues from the University of Montana made one of the most carefully documented studies of this serious problem. For a 30-month period between 1974 and 1976, they analyzed all sheep mortalities on Eight Mile Ranch, located about 14 miles south of Missoula. In spite of the fact that moderate coyote control efforts were made, these canines killed 1,223 ewes and lambs out of a total 8,459 pastured animals, a 14 percent loss. Not all sheep operations suffer such high mortality, but it is easy to see why some woolgrowers are getting out of the business.

Coyotes hunt at dawn or dusk and during the few hours following sunset, traveling over an area of from five to 40 square miles or more in search of food. While hunting, their keen eyesight, perhaps an adaption to the open terrain they always have inhabited, is the most important sense. Although coyotes are excellent runners capable of maintaining speeds of 25 to 30 miles per hour for long periods and reaching 40 miles per hour in short sprints, they seldom make lengthy pursuits of prey. They prefer to ambush small game with quick leaps, or make short rushes at larger quarry in the open.

Many coyotes pair for life. Their breeding season extends from late January through early April. Toward the end of the two-month gestation period, the female prepares several den sites, perhaps an abandoned badger or fox den that she remodels to suit her needs. Hollow trees, logs and crevices among rock piles may be used. Eventually the female selects one as a nursery where she gives birth to an average five to six pups, the number depending on population density and food availability.

Although white man always has persecuted coyotes he also, unwittingly, has helped them to survive. Not only did he open up forests previously too dense for their liking, he drastically reduced the numbers of their natural predators. Cougars, wolves and bears no longer pose a serious threat to the coyote. Possibly the highly adaptive song dog of the Northwest will outlive his main competitor, man.

Gray Wolf

When Lewis and Clark traveled through the Northwest in 1805-1806, they recorded that wolves were nearly everywhere, especially in the prairie foothills of the Rockies where bison were abundant. While in Fort Clatsop at the mouth of the Columbia, Meriwether Lewis noted wolves also were a familiar sight in the wooded country along the Columbia River, from the Pacific eastward through the Cascades. Throughout the explorers' travels, wolves hovered near the party's camp sites, quick to remove anything edible left behind.

Just a few years later, John Day (of fossil bed fame) and his partner Ramsay Crooks were saved, rather than harassed, by a wolf. Day and Crooks had been left by Astorian trappers to spend the winter of 1811-12 with the Snake Indians in western Idaho and northeastern Oregon. When the Indians moved on, the two starving visitors survived only because they were able to shoot a wolf that had ventured near camp. They immediately ate the skin, made broth out of the crushed bones and consumed the dried meat over the next several weeks.

Within 30 years, settlers followed these early explorers into the Northwest and the wolf (*Canis lupus*) was in for a hard time. White men harbored Old World notions of wolves as evil creatures out to dine on small children and weary travelers, and the fact that some wolves developed a taste for domestic livestock did not help their reputation.

With their inability to avoid man's wrath, it is easy to see how wolves were quickly exterminated in the Northwest, just as they had been elsewhere. Elimination of predatory wolves was a high priority in pioneering agricultural communities. At the Hudson's Bay Company's Fort Vancouver, on the site of today's Vancouver, Washington, raising livestock was an important adjunct to the fur trade. In December 1839 Chief Factor John

Right: The much maligned gray wolf now is extremely rare in the Northwest. Alan Carey photo.

McLoughlin received the following correspondence direct from the company's London headquarters, "We send by this conveyance . . . a small quantity of Strychnine made up in dozes for the destruction of Wolves; it should be inserted in pieces of raw meat placed in such situations that shepherd's dogs may not have access to them, and the natives should be encouraged by high prices for the skins to destroy wolves all seasons." Use of strychnine continued, and by the early 1850s a naturalist employed in the Northern Railroad Survey could report that wolves were "quite scarce" in the nearby Puget Sound Lowland.

To the south across the Columbia River, the need to subdue predatory wolves even was linked to early efforts to establish a provisional government in Oregon. As pioneering settlers moved into the Willamette Valley in increasing numbers in the 1840s, their livestock became easy prey to wolves lured to the area by the tasty domestic animals. In 1843 a series of meetings, which have passed into history as the Wolf Meetings, were held. They resulted in a full-fledged scalp bounty law. Large wolves fetched a price of $3.00, provided the skin of the head with ears intact was presented.

Granville Stuart, pioneer Montana cattle baron, noted that when beaver began to disappear from the Northwest, trappers turned to wolves, especially when their pelts increased in value in the mid-1860s. In the following decades, professional "wolfers" became something of a legend. Yankees, halfbreeds and others with suspect backgrounds led independent lives in barren country replete with hardships and dangers. Enough money to buy a modest ranch could be earned by selling pelts and collecting bounties from one winter's work, but often the small fortune was squandered on more immediate needs, like cheap whiskey.

When the U.S. Forest Service began collecting grazing fees on national forests, stockmen felt it was only fair that the federal government should also engage in predator control. Accordingly, forest rangers got into the wolf-trapping business in the early 1900s. Further federal involvement came in 1915, when the U.S. Bureau of Biological Survey began organizing western predator control districts, complete with supervisors and federal predator control agents. By 1939 most of the few remaining wolves in the Northwest lived on National Forest lands, but their numbers were declining rapidly. At that time less than 80 wolves were thought to remain in Montana and Idaho, while an estimated 140 continued to roam the wildest sections of the Washington and Oregon Cascades.

The Northwestern states were made "safe from wolves" by the 1940s, but slowly the pendulum of public attitude had begun to swing the other way. Credit for helping change public perceptions of the wolf is due Jack London, an author almost completely ignored by most modern-day wolf experts. London wrote such epics as *White Fang* and *The Call of the Wild,* along with many short stories featuring wolves as important characters. The Jack London Society reports that he is one of the most widely read authors in the world, and surely there are few American school libraries that do not contain several thumb-worn copies of his works. This author of adventure tales for youths and adults wrote about wolves and wolf-dogs from his experiences in the Klondike and the Northwest.

His stories were not based on scientific fact, and he actually helped perpetuate the myth that wolves are dangerous to humans, but he did show readers that wolves are worthy of respect in their wilderness role, and that there is a fine line separating the wolf from man's best friend, the dog. Anyone with even a passing interest in the outdoors can hardly help but have admiration for the wolf after reading London's works.

Little was known about the true habits of North American wolves until the 1930s and '40s, but today there are dozens of books dealing with the wolf, and much of this predator's basic biology has been learned.

Adult wolves ordinarily mate during February through April, with only one pair per pack usually being successful breeders. The female digs a den several feet long in which to give birth to three to seven pups. Both parents, along with other pack members, feed the pups by regurgitating food the adults have previously consumed. Later, whole parts of prey animals are brought to the den and given to the litter. At eight to ten weeks, the pups

In the late 1800s vagabond wolfers became something of a legend. L. A. Huffman, Montana Historical Society photo.

Right: Large prey animals are essential to the survival of the wolf pack. Bruce Pitcher photo.

leave the den and move to what are called rendezvous sites, small openings in the forest where adults continue to bring food to the pups for several more weeks. As the pups mature, they travel more widely with the pack and begin to hunt.

Pups are able to fend for themselves at one year of age, but they often stay with the parents, which seems to form the nucleus of many packs. Packs of eight or fewer animals are most typical. A "lone wolf" is relatively uncommon.

Adult males usually weigh about 85 pounds, and females somewhat less. A pack of such good-sized predators means that large prey animals are essential to the group's survival. Bison were the preferred food of prairie wolves, but today's rely on other traditional staples, such as deer, moose, elk, mountain sheep and even beavers. A pack of wolves may travel over an area from 50 to 5,000 square miles or more in search of prey.

The relatively recent accumulation of factual knowledge about the wolf has stimulated efforts to preserve the few remaining wolves. National interest led to the 1973 federal classification of the wolf as a threatened or endangered species in the lower 48 states. During the last few years, to the delight of conservationists and the wrath of stockmen, a federal-state Northern Rocky Mountain Wolf Recovery Team has been working to enhance wolf populations in northeastern Washington, northwestern Montana, northern and central Idaho and around Yellowstone National Park. It may be that wolves can be established in wilderness areas remote from livestock interests, but this is a hot political issue, as well as a difficult biological one.

Unless recovery efforts are successful, the chances of seeing a wild wolf in the Northwest are slim. Jerry Hickman, Regional Nongame Biologist with the Washington Game Department, reported there have been fewer than 10 reliable wolf sightings in the North Cascades in the last 10 years. A few occasionally drift down from Canada into Montana's Glacier Park area, northern Idaho and northeastern Washington, while possibly a dozen live in the River of No Return Wilderness of Idaho, according to James Gore, Endangered Species Team Leader with the U.S. Fish and Wildlife Service Boise Field Office. Many of the Northwest "sightings" of wolves are observations of the smaller coyote. Identification is made more difficult by the fact that wolves can have very doglike black or white coats, although a grizzled gray is most common. As a general rule, if you see a wolf-like canid noticeably smaller than a good-size German shepherd, it is probably a coyote. If it is bigger and stockier than a shepherd, it *may* be a wolf.

Red Fox

The wily red fox helps teach us a lesson about mammal classification. The rule to be learned is that the external colors of a mammal are not always reliable guides when identifying different species. Instead, the careful observer bases such determinations on more stable, hard evidence, like the number and shape of teeth, or the shapes of the cranial bones.

Despite the "red" adjective this fox comes in a variety of colors, leading early naturalists, including Lewis and Clark, to believe they were seeing different species. The majority of red foxes boast a luxurious, soft reddish-brown pelage, which sometimes acquires an attractive golden tone. The long, bushy tail is white-tipped. Underparts are white, while the legs and back of ears are black.

Occasionally other color variations occur, especially in western Washington where part of the present-day red fox population originated from individuals released from fox farms. A cross phase has a dark cross-shaped pattern across the shoulders and down the back, contrasting with the overall brownish-yellow coat. Silver-tipped hairs mixed with pure black ones create a highly sought-after form called the silver fox. The rarest color phase in the Northwest is the black fox, which is sometimes encountered in the mountains of northern Idaho and in the Cascades. These and other types usually share one obvious characteristic, the white tail tip.

Red foxes (*Vulpes vulpes*) are doglike mammals with sharp noses and large, erect ears. They have a slender body about three feet long and weigh between six and 15 pounds.

These carnivores inhabit environments ranging from wet coastal lowlands to frigid high alpine meadows. They prefer open, grassy terrain dotted with shrub thickets and small wooded patches, and often live near human settlements interspersed with forested areas. Their range covers most of the Northwest, with the major exceptions of the dry, desert-like areas of eastern Washington and the central and southeastern parts of Oregon.

Along the Oregon coast they are largely absent south of Newport, but ongoing logging in this region is likely to encourage them to expand their range farther south.

A mated pair settles within a home range similar to the one to five square miles claimed by individuals. A den is constructed, most commonly by renovating a marmot or badger burrow system, and five to six young arrive after a 53-day gestation period. Within several weeks the frolicking and energetic kits venture out, guarded by the watchful vixen, whose sharp warning bark at any sign of danger sends them scurrying back to the burrow. The suspicious nature of red foxes is evidenced by the female relocating her litter at least once, even if there are no disturbances at the original site.

Young foxes grow rapidly and normally leave the den after four months, but remain within the home territory for another two months, sharpening their hunting skills. When young are about six months old, the family disperses, with some of the young males possibly moving as far as 100 miles away.

Even where abundant, red foxes are difficult to observe because of their shyness and non-daylight activity periods. They spend most of the day sequestered on an open hillside with good visibility. Even in winter they seldom seek shelter from the weather, opting instead to curl up in the snow, covering their noses and paws with their fluffy tails.

These small canines usually are beneficial to humans. In the summer and fall, such agricultural pests as grasshoppers, mice, pocket gophers and ground squirrels are eaten. During winter, mice, voles, rabbits and hares are staples.

Predators of red foxes include the rare wolf, wolverine, fisher, lynx and, especially, the coyote. Occasionally a golden eagle may attempt to capture one, but usually with limited success. In recent years the fox's importance as a furbearer has increased greatly. Pelts that trappers hardly could give away in the late 1950s now bring $30 to $50 each.

Right: Red fox, Audubon lithograph. Audubon, Quadrupeds. *Photo from Special Collections Division, University of Washington Libraries.*

Below: The cross phase, with dark area over shoulders and back, is one of the red fox's most attractive color variations. Terry Lonner photo.

Left: The red fox. Richard E. Kirchner photo.

Black Bear

An unscheduled visit by a black bear is always a possibility when camping in the Northwest backcountry. Alan Carey photo.

Hungry black bears are nature's magicians. Friends have reported seeing determined animals hand-walking along a rope strung between two tall trees to reach a grub sack hanging in midair. Others have reported bears unscrewing the lid off a peanut butter jar, and even opening gum wrappers with unbelievably dexterous paws. Even water hazards are no obstacle to a determined bear. The wise canoeist need not be reminded that camping on an island that was bear-free before dark does not insure that a bruin will not visit before morning.

When not raiding campgrounds, black bears (*Ursus americanus*) naturally prefer relatively open forests with a shrubby understory, interspersed with numerous small meadows. Brushy thickets that quickly cover clear-cuts left by logging also seem to be to their liking. These bears might turn up almost anywhere in the Northwest, but typically are found in sparsely settled mountainous regions. Resident populations are absent only from the open plains of northcentral and southeastern Oregon, eastern Washington and southern Idaho.

Despite their name, close to half the black bears in the Northwest are some shade of brown. These are heavily built animals that appear almost comical when wandering through the woods with their head swinging from side to side. Their clumsy and slow appearance is deceiving, however, since the bear is capable of loping as fast as 25 miles per hour for short distances when necessary.

The bruins may measure up to six feet from the tip of the nose to the end of their stubby tail and stand three feet high at the shoulder. Males weigh between 200 and 300 pounds, though an occasional heavyweight may tip the scales at 600 pounds. Startled backcountry hikers happening upon extra-large, brown phase black bears commonly mistake them for grizzlies. Recent work by Dean Wheeler, wildlife biologist with the Oregon Department of Fish and Wildlife, suggests there is little chance of confusion in the Coast Range since black bears there average just 100 pounds.

Black bears are solitary, except for sows with their cubs, and pairs during the breeding season. Most of their time is spent rustling food within the home range. Boars roam over a domain of two to 45 square miles, while sows confine their movements to a significantly smaller area. Individuals avoid each other under normal conditions and, upon meeting, most commonly pass without paying much attention to one another.

Twilight or just before sunrise are the best times to find black bears out and about, though they may be seen foraging throughout the day, especially in spring and fall. During hot summer days they may spend considerable time wallowing in a cool pool of water to avoid the heat. At dusk their search for anything edible begins, guided by their well developed sense of smell. Although classified as carnivores, they eat far more vegetative materials than animal matter. Most time is spent grazing on succulent plants, such as newly emerged grasses and forbs. Various roots, nuts and especially, berries also are included in the diet. The bruin gladly eats any small mammals it stirs up in the process, but usually does not actively hunt them. Much of the animal matter consumed consists of insects, particularly those living in colonies. A bear finding an anthill breaks it with one strike of its paw and then lays the paw in the center of the chaos just created. After the defending insects have swarmed on the paw in sufficiently large numbers, the bear licks them off and repeats the act. In spring these mammals may rely heavily on carrion, or even kill large ungulates, like elk and deer weakened by the long winter. They also may prey on deer fawns or elk calves, especially if green forage is not abundant. They occasionally attempt fishing, often with poor results.

Black bears are dormant during winter. This requires that they accumulate large enough energy reserves to make it through five to seven months of fasting. In the fall bears are highly active and may travel up to 100 miles in search of a good meal. In Montana they have been found to gain between one and two pounds daily from August through October, adding 20 to 30 percent to their spring weight. The reserve is stored as a layer of fat up to four inches thick beneath the skin. With arrival of cool weather bears investigate several suitable-looking caves, crevices, hollow stumps or

Although classified as carnivores, black bears eat far more vegetative material than animal matter. Despite their name, close to half the black bears in the Northwest are a shade of brown. Alan Carey photo.

similar sheltered places for a den. As the food supply decreases they spend more time in the selected site. After a few days of fasting in late October or early November, the bear crawls into its den. In many areas of the Northwest falling snow both conceals and insulates the hideout.

After denning up, the bear becomes lethargic due to a slowdown in most bodily functions. Within the first three weeks its heart beat drops from a summer rate of 40 to 50 per minute to just eight per minute. Breathing becomes shallow, decreasing the oxygen intake as much as 50 percent. Metabolism is slowed down 50 to 60 percent and body temperature drops well below that of an active bear. Biologists disagree whether a bear's winter sleep should be called true hibernation since its body temperature does not drop as low as most

other hibernators, nor does it sleep as soundly. Some classify the bear as an official hibernator, just like marmots and ground squirrels. Others prefer to call the cold-season slumber "winter sleep" or "partial hibernation." Black bears, especially adult males, may emerge for short periods during winter warm spells and even change the den site. The lengthening days of spring usually bring the bears out for good in April or early May in most of the mountainous Northwest. In the warmer coastal regions they may emerge as early as late February.

The first weeks of spring are spent partly foraging and partly in total inactivity inside the den. Beginning in June the females without cubs are ready to breed, causing fights between rival males. Bears have delayed implantation, which means that after fertilization the embryos initially develop only slightly into a tiny ball of cells scarcely visible to the naked eye. The embryos then stop developing for about five months. Helpless cubs are born in late January or early February while the mother is still denned up. Early Northwestern naturalists were reluctant to believe the tales of old bear hunters, who told them they never could find a pregnant female. There was much validity to this observation, since only about one month of the seven-month gestation period is required for converting a small mass of cells into an identifiable bear cub fetus.

A sow's first litter usually consists of a single cub, while subsequent litters are commonly twins. The young ones are tiny, weighing six to 10 ounces, hairless and do not open their eyes until five weeks old. When they leave the den with their mother they weigh about four pounds. Emerging cubs are capable of following the sow, and then feed on the same succulents, although milk is still preferred. While foraging the sow carefully watches her offspring and sends them up a tree at the first sign of danger. With the onset of winter cubs den up with their mother, who weans them just before hibernation. The following spring the youngsters disperse, sometimes staying together for another year and slowly expanding their travels outside their familiar home range.

Garbage bears that are repeat offenders in national parks often are destroyed. Although begging bears are very appealing, people who feed them in parks are signing their death warrants. Bears accustomed to humans even can become a nuisance in towns. Problems with raiding bears became so serious during 1984 in West Yellowstone, Montana, that city fathers passed an ordinance requiring all private and public garbage containers be bear-proof.

Regionwide the black bear is a valued game animal with approximately 20,000 taken each year by hunters, three-fourths from western Montana. On the Olympic Peninsula and in the Puget Lowland special spring bear hunts are used to thin the population of bears that cause considerable damage to tree farms. Bears strip and eat the cambium, or inner bark of tree trunks, especially the commercially important Douglas fir, and may kill as many as 200 trees per acre.

In timber dependent Oregon it was statewide open season on black bears with no bag limit until the early 1960's. Residents of the western Washington timber town of McCleary thought they needed to do their part to help limit bear damage to local forests. The annual Bear Festival has been *the* community event each year since 1958. Held the third weekend of July, the festivities include a bear stew feed in the city park. Despite demonstrations by Seattle bear preservationists, the annual Bear Festival remains the high point in McCleary's social calendar.

Above: Black bears are not as clumsy as they look. Not only can they move along at speeds up to 25 miles per hour, they also can climb trees. Alan Carey photo Below: Pacific Coast black bear cubs nap on big-leaf maple. Thomas W. Kitchin photo Left: Welcome to McCleary. John A. Alwin photo.

Grizzly Bear

Outdoor magazine writers and television show producers are fond of characterizing the grizzly bear as the largest and most ferocious terrestrial carnivore south of the Canadian border. Usually overlooked is the much less dramatic fact that, depending on season and availability, 50 to 100 percent of a grizzly's diet consists of items of which the most devoted vegetarian would approve.

This bear's reputation as a meat eater probably has been enhanced by its activities during the early spring, when it may move to lower elevations where large ungulates have wintered. Bears are more easily observed at this season than at any other. Grizzlies (*Ursus arctos*) emerging from dens in the spring have high energy demands due to their considerable weight loss during winter hibernation. Such needs may be met by scavenging carcasses of deer and elk as well as by killing some of those weakened by winter malnourishment. Before long, however, the meadows begin to green up and bears spend as many as 20 hours per day grazing on newly emerged vegetation.

As spring progresses, bruins may follow elk herds to their calving grounds and feed on newborns. Soon these become too quick for the large carnivores to capture, and the bears begin their slow migration to higher elevations on the heels of the receding snow. Various succulent plants, tubers and bulbs make up the bulk of the diet by midsummer. Burrowing rodents, including squirrels and marmots are prized delicacies and the bear may spend considerable time and effort digging through the meadows in pursuit of these tasty tidbits.

During the latter part of summer and early fall, grizzlies must accumulate an adequate layer of fat, up to eight inches thick and weighing as much as 400 pounds, to see them through the approaching winter. This can be achieved by consuming large quantities of huckleberries, grouse whortleberries and serviceberries. In years with good berry crops, they may be the bear's only food for

Right: Montana's Northern Rockies are home to the largest population of grizzlies south of Canada. Thomas W. Kitchin photo.

weeks—hardly consistent with the animal's blood-thirsty image. In fall, pine nuts are another excellent source of energy in many parts of their range. Grizzlies especially savor those gathered by red squirrels from whitebark and limber pines.

Grizzlies' solitary lives are spent within a well-defined home range, the size of which depends on the animal's age and sex, as well as the quality and abundance of food sources. Within Yellowstone National Park, for example, ranges vary from about 20 to more than 1,000 square miles. Bears move constantly about the various parts, concentrating on each food source as it becomes available. Grizzlies evidently do not defend their home ranges, and most are overlapping. As many as 23 animals have been observed peacefully congregated around a ripe carcass. After the largest beast has eaten its fill, the others get a chance according to the feeding order.

Bears breed from June through early August. Usually the female is courted by just one male, but if two boars have eyes for the same sow, fights may erupt. Two or three cubs arrive in February or March, helpless, blind and weighing a mere pound. After two months of nursing they emerge from the den as hefty 10-pounders, accompanied by an extremely protective mother.

Cubs spend the summer and fall with the sow, learning to distinguish between edibles and inedibles, and with the onset of winter den with the female. Most mothers remain with their cubs for another year, weaning them as two-year-olds. Thus, the majority of females have offspring only once every three years and a few may wait as long as four or five years between litters. Such low rates of reproduction mean the loss of just one breeding female in a small, isolated population can be critical.

"Old Ephraim," as the grizzly sometimes is called, spends winter hibernating in the same manner as the black bear. Most grizzlies prefer to dig their own dens, but will settle for natural cavities. To conserve body heat the den is made just big enough to accommodate the bear. Pregnant females are the first to retire for the winter, usually at the beginning of November. Others follow suit within a few weeks. Most leave the den and begin their search for food in April, while a sow and her cubs do not emerge until early May.

A grizzly bear's fur coat varies from dark brown to a medium tan. Guard hairs are white or silver tipped, giving the animal its grizzled look. Silver-tip females weigh about 300 pounds and males more than 500. The heaviest individual in one Yellowstone National Park study reached a formidable 1,120 pounds. They measure up to 6-1/2 feet in total length and three feet-plus at the shoulder. The more massive head, dished face, distinctive shoulder hump, larger size and often grizzled look are useful clues in distinguishing them from their black relatives.

At the beginning of the 19th century, grizzly bears could be found almost anywhere in the Northwest. They apparently were abundant along major river valleys in Montana and Idaho, but were seen less frequently west of the Cascades. Proof of their presence in Oregon's Willamette Valley is found in the journal of noted botanist, David Douglas (after whom the Douglas fir is named). While traveling through the valley in the fall of 1826 a member of the Douglas party narrowly escaped death when a large male grizzly chased him up a nearby oak. That bear managed to elude hunters, but 20 years earlier the Lewis and Clark Expedition already had foretold what the Northwest grizzly's fortunes would be under the onslaught of the white man and his firearms. While just "passing through," this small band of men killed 43 grizzlies, sometimes for food or in self-defense, but often just for excitement.

The pioneering attitude toward troublesome grizzlies is obvious in a written account by early 19th-century Astorian trapper John Day, who admonished a young partner, ". . . caution is caution, but one must not put up with too much, even from a bear. Would you have me suffer myself to be bullied all day by a varmint?"

Despite decades of year-round persecution by hunters, trappers and thrill seekers, Northwest grizzlies managed to hang on in fair numbers until the early 20th century. As late as 1883, hunter-author William H. Wright reported seeing 11 grizzlies on a weekend hunting trip near Spokane.

With the quickened pace of development in the 20th century, the grizzly's days were numbered. The last reported grizzly killed in Oregon was shot in the Chesnimnus Creek area of Wallowa County

in 1931 by a government trapper. The last known kill of a Washington grizzly occurred in the North Cascades in 1967. Idaho gave complete protection to its few remaining grizzlies in 1947, but Montana had enough bears to permit fall hunts during the '50s and '60s, with about 40 bears killed annually.

Public pressure as part of the environmental movement of the '60s and '70s led to classification of the Great Bear as a threatened species in 1975. Still, hunting of grizzlies in Montana's northern continental divide area continued, with an annual mortality of 25 bears permitted from all human causes.

With legal hunting eliminated as a major population factor, biologists have concentrated their efforts on delimiting and preserving prime grizzly bear habitat and safeguarding established bear populations. According to Dr. Chris Servheen, Grizzly Bear Recovery Coordinator with the U.S. Fish and Wildlife Service, estimates of the current minimum numbers of bears in various Northwest areas include Glacier Park and nearby forests, 200; Yellowstone Park and adjacent sections of Montana and Idaho, 200; Bob Marshall Wilderness Complex, Montana, 380; and Selkirk Mountains of northwest Idaho and northeast Washington, 15. Less than 10 bears are thought to exist in each of the following general areas: Cabinet Mountains, Montana; Purcell Mountains, Montana; North Cascades (Okanogan National Forest), Washington.

Grizzlies have been able to survive in the Northwest because they could retreat to large, remote and uninhabited mountain forests. Now recreationists even follow the bears to these few remaining secluded places. A particularly troublesome sort of recreation from the big bear's perspective is the large commercial downhill ski complex, with its attendant roads, condominiums, homes, and hordes of visitors. Biologists and bear watchers especially are concerned when potential ski resorts are in and adjacent to prime grizzly range. Even talk about such resorts near West Yellowstone, Montana, and in the Selkirk Mountains of north Idaho raises the ire of the bears' loyal supporters.

Not everyone is enthusiastic about having more grizzlies in this world. Some bruins have been

Above: Grizzlies can be distinguished from brown-phase black bears by their more massive head, dished face, distinctive shoulder hump and often grizzled look. Thomas W. Kitchin photo Above right: Bronze grizzly, symbol of the University of Montana and its athletic teams, is displayed prominently in the central oval on the Missoula campus. Sculpted by Rudy Audio. John A. Alwin photo Far right: Presence of grizzlies in Glacier National Park means extra precautions are in order within some sections of the park. John A. Alwin photo Right: Young cubs like this one remain with Mom for two years. Alan Carey photo.

Restriction

Only Hard Sided Trailers
Or Recreation Vehicles
No Tents, Tent Trailers
And
No Sleeping On Ground

Reward poster shows the seriousness with which the National Audubon Society views the threatened grizzly bear. National Audubon Society.

notorious livestock killers and, of course, repeated grizzly attacks on humans in Glacier National Park always are given full coverage by regional and national news media. Gruesome tales of bear attacks in Jack Olsen's popular *Night of the Grizzly* have not helped the bruin's public image. Concerns are not alleviated by reminding people that the odds of a park visitor being injured by a grizzly were calculated to be two million to one, whereas the injury risk of driving a car through the park was much greater. The public's inordinate fear of grizzlies was apparent in 1970, when the Idaho Fish and Game Department briefly considered accepting bears from Yellowstone National Park for a restocking effort in Idaho's Selkirk Mountains. Letters of opposition poured into state and federal offices, and the project quickly was shelved.

Raccoon

In our region almost any lowland stream has what looks like miniature human footprints along the shoreline. The tracks follow the bank for some distance, mysteriously disappear into the shallows, and then resume on shore a few yards farther along. The evidence was not left by "Littlefoot," but rather by a familiar grayish-brown, masked creature of the dark who displays a bushy tail adorned with six black rings.

During their nightly expeditions, raccoons (*Procyon lotor*) follow creeks and other waterways in search of a favorite prey—crayfish. After selecting a likely spot for dinner, the raccoon wades in and feels around the rocks with its extremely sensitive forepaws. It appears to pay little attention to goings on because it does not follow the progress with its eyes, but gazes dreamily into space.

The genus name for the raccoon, *Procyon*, means "before the dog." There has been some question about the origin of this name, but it likely refers to the well-known fact that coons give a pack of hounds a good chase. The species name, *lotor*, means "a washer." Since raccoons routinely submerge food items before eating, people believed they were washing their food. Even today, the common name for the raccoon in such languages as Finnish and German is "washing bear." Rather than being fastidious creatures, it seems more likely that wetting their front paws is the objective of the behavior. The wetness probably enhances the sense of touch, making it easier to discard any

Black mask and ringed tail, raccoons are easily identified. Bruce Pitcher photo.

inedible parts of the meal. The "washing" might also reduce the amount of sand and minimize wear on teeth.

Even though classified as carnivores, raccoons are among the most omnivorous of animals. Accordingly, their diets include almost anything with calories. Frogs, worms, minnows, voles, mice and other small mammals, as well as birds' eggs and nestlings are all welcomed. Marine mussels, fish, shrimp and other animals frequenting the tidewater zone along the coast make up their entire menu in Washington's Willapa Harbor National Wildlife Refuge. In the fall, raccoon diets shift more toward vegetable matter, especially various berries.

The raccoon has earned his bandit's mask with distinction, having absconded with melons, sweet corn, apples and just about any ripening garden crop. Experienced trappers know that corn on the cob is an almost irresistible bait. In populated areas raccoons have learned to climb into trash cans and get their share of edible garbage. It appears that the raccoon's range has grown right along with the supply of garbage in the Northwest. In pioneer times, they were most abundant in coastal areas and the major river valleys, but rarely were seen in the interior. Now they are found everywhere except high mountains.

Raccoons are sedentary and generally solitary, although each home range may overlap another. When paths of two individuals cross, an impressive threat display ensues. Both animals bare their teeth and, while hissing and growling ferociously, raise their back hair. In most cases, this is enough to convince each to retrace his tracks.

In January, males begin their long search for mates and may travel several miles to find as many females as possible. After an approximate 65-day gestation period the single litter of the year is born, usually within a large leaf nest in a hollow tree. If the female is unable to locate such a tree, she will make do with a cavern under a wind-fallen tree, an abandoned fox den, or even a granary or hayloft. An average of four to five (two to three in western Washington and Oregon) offspring are born in April through June.

Each youngster must eat considerable amounts of feed since survival through the winter may depend on thickness of the fat layer deposited during fall. By the following spring more than half the weight gained the previous autumn may have been lost. The mother and her offspring usually remain together their first winter, during which time the active animals continuously search for food. Inclement weather may force them to remain inside for weeks at a time.

Some Indian tribes believed raccoons had great spiritual powers and were capable, if in a good mood, of making a man rich. They used the animals' pelts for robes. The meat was relished, and their highly valued fat was used as a hair dressing, as well as an ointment for bruises and wounds.

Early fur trappers in the Northwest had little interest in the raccoon, but this was to change. As the beaver trade waned in the 1830s, European tastes switched to raccoon. The 1840 to 1860 period was a golden age for raccoon pelts, which were favored by some Europeans for hats and garment trimmings. Scarcity of coons at that time meant the Northwest contributed little to meeting the demand. In 1855 the Hudson's Bay Company officer in charge at Fort Walla Walla noted that the Company collected a mere 2,000 annually. Today we have the coons, and Northwest trappers take 20,000 yearly, even with Davy Crockett hats passe'.

Marten and Fisher

A late-night traveler cruising down a deserted U.S. Highway 12 west of Lowell, Idaho might have to slam on his brakes to avoid adding a careless fisher to the traffic fatality list. Unless the motorist were a trapper, he probably would not realize that the terrier-size mammal he narrowly missed may be worth more money than a decent grade horse.

Dense, old-growth coniferous forests in the Rocky Mountain, Cascade and Coastal ranges provide ideal habitat for two of the most chic furbearers in the Northwest, the marten (*Martes americana*) and the fisher (*Martes pennanti*). Because of their thick and shiny winter pelages, both animals have long been prized by trappers. The marten, sometimes called the American sable, is susceptible to heavy trapping because its trusting nature and enormous curiosity cause it to readily enter a trap placed in a cubbyhole. Scarcity of the fisher makes the animal only more desirable, boosting prices to more than $300 for a dark, prime female pelt.

Neither species tolerates man's presence well, and encroaching human settlement in the Cascades of both Washington and Oregon has been one of the major reasons for reduced numbers in those areas. Alteration of their preferred habitat by logging has further decreased their ranges.

Highly prized by trappers, the fisher is now uncommon in the Northwest. Thomas W. Kitchin photo.

Today they primarily occur in the truly wild parts of the Northwest, including northcentral Idaho, Montana's Bob Marshall Wilderness Complex and in Glacier, Olympic and North Cascades national parks.

Like most members of the weasel family, both carnivores have short legs and long bodies with bushy tails. The marten is nearly the size of a skunk, about 26 inches long and one to three pounds. Its color varies from dark yellowish-brown to reddish-brown, with a light yellow to dark orange patch usually present on its throat. The fisher is nearly twice the size of its smaller relative, and weighs up to four times as much. It has a dark brown to blackish-brown body. The head, neck and shoulder areas are covered with white-tipped hairs, giving it a frosted appearance.

The marten is quite at home in trees. Tom and Pat Leeson photo.

52

Both are capable climbers, although the marten is more at ease with aboreal life. Its sharp, semi-retractable claws give excellent traction when chasing squirrels, which can be pursued almost everywhere. The long, bushy tail helps the marten regain its balance after making one of its incredible leaps (as long as eight feet) from tree to tree. Neither creature is averse to taking the express route from tree to ground, plummeting as far as 20 feet to a soft snow or pine needle landing.

These carnivores are primarily mammal eaters. Martens eat smaller prey, mainly rodents. They also are adept at catching red squirrels in the tree tops. When hunting gets tough carrion will be taken, and berries are acceptable belly-fillers. Fishers prefer snowshoe hares, but porcupines, mountain beavers or anything the marten eats also suit them. This "black fox" of Lewis and Clark is a skillful hunter of the formidable quill-covered porcupine, which may outweigh it by several times. Some long-time residents of the Elk City, Idaho area in the state's northcentral section think the early 1960s reintroduction of the fisher explains the area's lack of porcupines.

Martens and fishers both have an average of three offspring. Young martens are weaned at the age of six weeks, and they grow rapidly to adult size in just three months. Fishers, on the other hand, do not wean their young until they are 17 to 18 weeks old, at which time they begin pouncing on small prey. Young fishers also are slower growers than their cousins and do not attain full size during their first year. The low reproductive capacity of each species undoubtedly has contributed to the decline in their populations, since they cannot rapidly rebound from a severe loss.

Coyotes, cougars, wolverines and even great-horned owls prey on martens and fishers. Fishers, in turn, sometimes cannibalize their marten cousins. Northwest trappers take approximately 4,000 martens annually, with half coming from the old growth forests of Idaho. In the mid-1980s the much rarer fisher could be trapped legally in a five-county area of western Montana, where just 10 can be taken per season. The increased incidences of accidental catches in northcentral Idaho, approximately 30 per year, has led to a detailed study of the regional fisher population to see if limited trapping can be justified.

Long-Tailed Weasel

Long-tailed weasels can be found region-wide in most habitat types from sea level to alpine areas. These animals, along with their cousins, the short-tailed weasels, are among the Northwest's smallest carnivores, yet they are accomplished predators. They attack prey as large as a snowshoe hare, the equivalent of a terrier pulling down a mule deer.

Weasels have inaccurately been labeled blood-thirsty, cold-blooded murderers. The fact that they kill their prey by piercing its skull with their large canine teeth may have added to the small mammal's image as a savage. They have evolved into extremely efficient predators, using their serpentine bodies to the fullest when attempting to capture ground-dwelling rodents in their tunnels. If a weasel encounters an abundant food source, such as a squirrel nest full of young, it will take advantage of the situation, killing all youngsters and caching them in some nearby burrow for future use. Staples in the weasel's menu are various small rodents, such as voles, mice and shrews, but young cottontails and mountain beavers also are eaten. The diet commonly is supplemented with insects, birds and eggs.

Left: In snowy areas, long-tailed weasels turn white each winter. Only the tip of the tail remains black. Thomas W. Kitchin photo Right: Long-tailed weasel wears its summer coat in the Cascades. Thomas W. Kitchin photo.

Long-tailed weasels (*Mustela frenata*) are the largest of all weasels, their slender bodies between 12 and 14 inches long, of which half or more is black-tipped tail. Males are 10 to 15 percent larger than females, and weigh from three to nine ounces. Weasels' summer pelage is brown on top and yellowish-white underneath. In areas with frequent snowfall, the animals turn white during winter, except for the black tail tip. In warmer parts of the Northwest, such as west of the Coastal Range, the ground usually remains snow free and these mammals stay brown year-round.

Adults are solitary, avoiding contact with each other except during the breeding season. Each individual confines its movements within a 30- to 40-acre home range that varies in size according to habitat, season, food availability and population density. The territory is vigorously defended against intrusion by other individuals.

This diminutive carnivore has a 278-day gestation period, nearly the same length as that of the continent's largest herbivore, the bison! Due to delayed implantation the fetuses do most of their growing during the last 27 days of the pregnancy. The single litter of the year, consisting of six to nine helpless young, is born in April or May. Youngsters remain blind for five weeks, but they grow rapidly, attaining adult size in three to four months. Surprisingly, the little ones are capable of fending for themselves at just seven to eight weeks, when the family usually disperses.

These voracious predators may provide a meal for larger carnivores, such as foxes and coyotes. Great-horned owls, various hawks and even rattlesnakes eat them, making a weasel more than three years old a rarity.

European monarchs once placed great value on white ermine (weasel) skins as adornment for royal robes. The long-tailed weasel is sold on the fur market as ermine, although technically the true ermine is the short-tailed weasel. With an average pelt bringing a trapper little more than a dollar apiece in today's market, few trap this most widely distributed North American weasel.

Mink

The ancient close relationship between man and other mammals is nowhere more clearly demonstrated than in Indian legends passed from generation to generation. In these stories aboriginal people imbued mammals they knew well with human thoughts, acts and morals, and related remarkable insights about the natural habits of the mammals themselves. The legend of "Mink Kills Whale," told by the upper Chehalis Indians of western Washington, is an intriguing example of how a factual observation may have been turned to myth.

The story describes a mink's adventures while on a whaling expedition. The mink observed a whale passing near the coast, and paddled his boat out to his prey. The diminutive predator then cut a hole in the side of the behemoth and entered

Below: Mink, Audubon lithograph. Audubon, *Quadrupeds. Photo from Special Collections Division, University of Washington Libraries.*

its body. After ordering the whale to swim closer to mink's friends, he cut out the whale's heart, and the dead whale was towed to shore. Mink's friends did not believe the little animal could have killed such a big foe, until they saw him emerge from the hole in the whale's side.

Although it seems preposterous to think that a mink could go whaling as did humans, the story may have been based on knowledge of the mink's habits, as well as a real experience. First, it is well known today that the mink (*Mustela vison*) is a ferocious predator. It preys on animals several times its size, and sometimes kills much more than it can possibly eat in the foreseeable future. Secondly, minks are not above scavenging meat, and those living in wetlands along the coast would not mind gnawing on a beached whale, although they prefer fresh meat, as any mink farmer will verify. Thirdly, a close relative of the mink, the wolver-

ine, has been known to crawl inside a dead whale and feast for days. A mink likely would do the same, especially if it were trying to reach the highly preferred red meat beneath the blubber. An entrance could be made through some external wound. Now, if you were hacking at the side of a beached whale and a much disturbed mink suddenly emerged from an old wound in the whale's side, what might you think?

The central character in "Mink Kills Whale" is a sleek-bodied member of the weasel family that moves with graceful bounds. Its coat varies from uniform rich chocolate brown to almost black, usually with a white chin and a few scattered white spots decorating the throat. Adult males are about two feet long, of which the moderately bushy tail accounts for some eight inches. Males weigh between three and four pounds, with females seldom more than half this size.

Minks are common near Northwest rivers, creeks, lakes and marshes. Along the Pacific Coast they frequent the brackish waters of estuaries and salt marshes. They appear more tolerant of human presence than their larger relatives, the marten and fisher. Minks often thrive near cities, but they seldom are seen because of their nocturnal lifestyle. These agile carnivores usually build their dens in any protected place at the water's edge. Vacant bank burrows of muskrats and beavers often are recycled, as are hollow logs or muskrat houses. If nothing ready-made can be found, a mink will dig its own den in a stream bank.

Within their relatively large home ranges, approximately 20 acres for the females and up to 50 times that size for males, minks have several "activity centers." These areas are used more heavily by the resident animal than other parts of its range, although the entire home range is covered by the owner over a longer period of time. Each activity center has from three to five useful dens. When the food supply declines in the vicinity of one center, the animal moves on to the next.

Although these mammals are generally solitary, the large home ranges often overlap. Resident minks are extremely hostile toward any trespassers. Males fight viciously throughout the year to defend their turf, with battles sometimes ending in death. Mink hunting territories are marked with a foul-smelling discharge from the anal gland, which many humans consider at least as rank as that of skunks. The main reason for this behavior seems to be to attract members of the opposite sex. If the "perfume" works, a lucky male might mate with two or three females during one breeding season, which extends from the beginning of February through early April. Infrequently a male briefly shares a den with one of the females before returning to bachelordom, leaving the female to care for her two to 10 kits.

Minks are excellent swimmers and spend much of their time in water. Not surprisingly, aquatic animals make up the bulk of their diet. Prey includes muskrats, frogs, fish, turtles and marsh-dwelling birds like ducks. Coastal residents hunt for small crustaceans and fish in pools left by the receding tide. Though they seem to prefer a watery

54

environment, minks also are cunning hunters on land, adding rabbits, mice, chipmunks and snakes to their menu. Often, when the prey is bigger than what the animal can eat at once, the excess is stored in a den for later consumption. One cache was found to contain nine muskrats, five coots and four ducks!

Mink predators include foxes, bobcats and great-horned owls. Northwest trappers also prize mink, taking more than 9,000 annually. Many wild mink pelts are used for trim on coats and hats, but those luxurious mink coats usually come from animals raised on fur ranches. This industry reduces trapping pressure on our native mink, and helps assure that we will see the tracks of this little "whale-killer" on the shores of creeks and marshes for a long time to come.

Wolverine

A junkyard dog may be the "baddest" animal in town, but the wolverine is surely the meanest one in the woods. Few animals have generated as many legends as the elusive and mysterious wolverine. It has been represented as a fierce animal with extraordinary strength, agility and intelligence. Wolverines are said to have attacked every living creature on this continent excluding man, and are said to be capable of driving a bear away from a fresh kill. Attacks on mountain goats have been documented, and there are reliable reports that a wolverine can bring down a bull moose if the quarry is hindered by deep snow. More typical prey is likely to be beavers, marmots, porcupines, snowshoe hares and grouse, along with vegetable food like berries and roots.

Wolverines (*Gulo luscus*) are thought to have feeble eyesight, while their acute sense of smell sometimes leads them to tear into backcountry cabins to devour food stored inside. They also are accused of following trap lines and relieving them of bait as well as victims.

French Canadian fur traders prized what they called the *carcajou*, as did the Hudson's Bay Company, which referred to them as the "quickhatch."

Drawn on stone by R. Trembly

Wolverene

Drawn from Nature by J.J.Audubon.F.R.S.F.L.S.

Printed & Col.d by J.T Bowen, Philad.a

Left: Wolverine, Audubon lithograph. Audubon, Quadrupeds. *Photo from Special Collections Division, University of Washington Libraries Below: Wolverines typically prey on hare- and marmot-size mammals, but are capable of bringing down some big game animals. Alan Carey photo.*

Today, among the four Northwest states, only Montana allows trapping, and in recent years annual harvest there has ranged from six to 58 animals. In addition to river and sea otters, wolverine pelts rank a full "100" on the durability rating scale. The fur is highly valued as a trim around the hoods of parkas, although its frost resistance is merely myth. However, its durability keeps it from breaking off when frost is removed from the hair, making it last longer than other types of fur.

This largest land-dwelling member of the weasel family does not resemble such slender, fast moving creatures as martens and minks. It looks more like a streamlined little bear with a bushy tail. The dark brown coat often is adorned by two dirty yellow bands along each side that meet at the base of the tail. Adults weigh between 18 and 42 pounds and are reputed to be the most powerful mammal for their size.

Wolverines probably never have been abundant in the Northwest. Even the alert members of Lewis and Clark's party were unable to positively identify one. Nevertheless, small numbers of wolverines still inhabit the forests and alpine tundra of the high mountains in the Cascades and Northern Rockies. Extreme remoteness from man evidently is not the prime prerequisite for wolverine habitat. They are known to inhabit the heavily used Bridger Mountains just a few miles north of Bozeman, Montana.

Food is scarce in the wolverine's harsh habitat, forcing the animals to roam huge areas in search of prey. Home ranges of males may exceed 1,000 square miles and usually are shared with two or three adult females. Their mating season is exceptionally long, lasting from April through September, possibly to ensure that the animals will meet while able to breed.

Wolverines have few natural enemies. In earlier times a wolf pack may have captured a Northwest wolverine if no trees were around for a retreat. Even trapping probably is not too hard on wolverine populations since few trappers would be foolish enough to count on catching such a wide-ranging, scarce animal. At least in the Northwest, most traps used for forest-dwelling furbearers like marten and fisher are not strong enough to hold the determined "weasel-bear."

A formidable array of claws makes the badger a world-class digger. Bruce Pitcher photo.

Badger

Badgers work hard for a living, since they have to dig faster than the various small ground-dwelling mammals which constitute the bulk of their diet. These medium-size carnivores are well-adapted for the task, having feet equipped with long, recurved foreclaws for digging, and shorter shovel-like hind claws to send the dirt flying. Badgers literally can dig themselves out of sight in minutes in sandy soil. They are so quick that a team of 10 men equipped with shovels could not unearth one given a mere one-minute head start.

French trappers working for the North West Company knew the "brarow" well, long before Lewis and Clark first saw one in North Dakota early in 1805. The explorers' written descriptions of these 14- to 25-pound mammals were new to science, so they were credited with its "discovery" in North America.

Badger pelts are shaggy and a grizzled gray with a striking white stripe running from snout to shoulders. They are prized by fur trappers who harvest 3,000-plus in the region each year, more than half coming from western Montana. The fur is especially well-suited for the manufacture of shaving and other types of brushes.

Badgers (*Taxidea taxus*) are largely solitary, except for females with young and short-lasting pairs during the breeding season. Adult animals spend their entire lives within the same home range. In southwestern Idaho ranges average 600 acres for adult males and 400 acres for grown females.

Several dens within their home range is standard for badgers. The entrance of any actively-used badger hole is fairly easy to detect because of the bones, fur and possibly rattlesnake rattles littering the ground around the opening. If the owner is home, the entrance may be plugged by loosely packed soil, especially during cold weather. In summer, badgers seldom use a den for more than one day at a time. During cold weather the animals become torpid, spending several days inside the same burrow.

The breeding season commences in July and extends through August. Due to a six-month delayed implantation, the well-furred, blind young are not born until the following spring. Youngsters suckle for five to six weeks, but the female also feeds them solid food. Following weaning they remain with their mother for another month and a half before dispersing.

Badgers inhabit open terrain covered by grass or sagebrush, from sea level to high-mountain meadows. Their range includes most of the suitable habitat in the Northwest, except for wetter coastal portions of Oregon and Washington. They spend their days dozing in their dens, and nights hunting for small mammals, including ground squirrels, pocket gophers, mice and voles. If a badger accidently stumbles upon carrion, birds or even insects it will gladly eat these, too. Many people might be happy to know that badgers also are fond of fresh rattlesnake!

Destruction of habitat due to agriculture may be the greatest threat to these mammals in the Northwest. Many ranchers do not appreciate them, even if badgers consume large amounts of agricultural pests, because they consider badger holes hazardous to livestock. The courageous badger is a formidable opponent for any predator, even mountain lions. It quickly retreats into a burrow, or digs a new one, from which it can bite and claw at an adversary while snarling, growling and hissing—all of which are intimidating.

Sweet and sour, Pacific bleeding hearts and striped skunk. Thomas W. Kitchin photo.

Striped Skunk and Spotted Skunk

Maybe it is too bad that our large supermarkets do not stock fresh skunk in their meat counters. A botanist traveling in 1810 with the Wilson Hunt and Ramsey Crooks expedition reported that skunk meat was considered a delicacy by some members of the party. This observation was convincingly affirmed by a naturalist writing in 1860, "A large fat skunk, carefully prepared, I saw cooking in a camp on the Blue mountains of Oregon. . . . I, finding the creature so much to my fancy, made a hearty dinner of it. . . . they are certainly good eating; the slightly strong flavor resembling much that artificially given by a skillful *chef de cuisine* with onions or garlic." Evidently American tastes just aren't what they used to be!

They cannot be found in a grocery store, but the common striped skunk (*Mephitis mephitis*) occurs just about anywhere else in the Northwest where there are a few brushy stream bottoms and possibly a hen house or two to raid. The smaller spotted skunk (*Spilogale putorius*) prefers similar habitats, but is generally much scarcer, and is absent in large sections of the northern Rockies and much of eastern Washington.

The species names for both animals, *putorius* and *mephitis,* come from Latin words meaning "a foul odor," and these mammals exhibit one of the most effective chemical defense systems in the animal kingdom. Each has two small musk glands embedded in the muscle on either side of the tail. The powerful musk usually is not immediately used when a potential predator approaches. Instead, the skunk gives fair warning. It may stomp its front feet, hiss, growl, arch its back or raise its tail. The display may include an acrobatic walk on the front legs. If all this fails, the skunk quickly bends into a U-shaped posture, with bushy tail

raised above an arched back and head between hind legs to keep an eye on the enemy. The next instant, one or two streams of the devastating, skin-burning liquid is fired with astonishing accuracy to distances up to 20 feet. The performance can be repeated as many as six times, if necessary, before time is required for reloading.

While this defense has proven satisfactory with mammalian predators, it is not so useful against those equipped with wings or four wheels. Every spring and summer the roads in rural areas are dotted with foul-smelling carcasses, and one of the sure signs of an active great-horned owl nest is the pile of skunk skeletons, 57 of them in one case, scattered on the ground below the tree.

Despite their formidable armament, skunks are peace-loving creatures. Male striped skunks try to avoid one another during the summer, but they may fight valiantly with lots of squalling, growling and hissing for the females during the late February through March breeding season. Individuals of the spotted variety are even more social, often sharing the dens in one area with all members of the community.

Skunks often renovate abandoned coyote, fox or even mountain beaver dens to suit their style. They also will use natural crevices in rock piles, hollow trees and logs. If necessary, they excavate their own dens, sometimes under the porch of an unfortunate farmer. In summer dens rarely are used for more than one week at a time. However, during winter skunks den up for as long as three months. They do not enter true hibernation during cold weather, but are drowsy and lethargic, living mainly from fat built up in summer and fall. Each thaw finds the skunks, especially adult males, up and about.

Although closely related to notorious meat eaters like minks and weasels, skunks are more omnivorous. Especially in fall and winter, berries, plant roots and seeds are important, and may make up half the diet. Year-round, nightly hunting expeditions in meadows and fields yield animal foods, such as mice, duck eggs, grasshoppers, beetles, potato bugs, snails and grubs. Many of these prime skunk foods are agricultural pests. Crayfish, earthworms, amphibians, carrion and birds (in the egg or out) are especially favored by

Ready, aim, . . . Bruce Pitcher photo.

the more agile spotted skunk, who is quite at home climbing trees. Many skunks are attracted to human dwellings, where there is always a chance to find garbage or livestock feed.

Most intentional trapping of skunks today is to eliminate nuisance animals around buildings or, more rarely, to help control rabies outbreaks. This was not the case in years past. In the 1860s thousands of skunk pelts were shipped to Great Britain annually, and the price of a skin rose to $1.00 in the 1870 to 1890 period, more than $30 apiece in current inflated dollars. Unfortunately, a modern trapper receives less than $3.00 for the unpleasant task of preparing an odiferous skunk pelt for market.

River Otter

"The most eligant piece of Indian dress I ever saw," was the way Meriwether Lewis characterized the ceremonial fur "tippet" worn as a neckpiece by the Shoshone (Snake) Indians. The tippet was made from a several-foot-long strip of dressed river otter skin bordered with 100 to 250 white, long-tailed weasel pelts. Lewis first described the neckpiece in August of 1805 while camped in eastern Idaho along the Lemhi River. He noted that the garments were highly esteemed by the Indians, and that they were presented as gifts only on important occasions. After Shoshone chief, Cameahwait, gave an otter tippet to Lewis in August of 1805, the explorer was able to preserve the garment until he returned to Philadelphia where artist Charles de St. Memin rendered his well-known drawing.

The reluctant donor of the main ingredient in a tippet is a highly aquatic member of the weasel family. River otters, the land otter of the fur traders, have elongate (35 to 51 inches), slender bodies, and weigh between 11 and 30 pounds. Their long, tapering tail accounts for almost one-third of the body length. Animals are covered by one of the most lustrous and durable furs of any mammal. It

River otters are especially well-adapted to their aquatic world. Alan Carey photo.

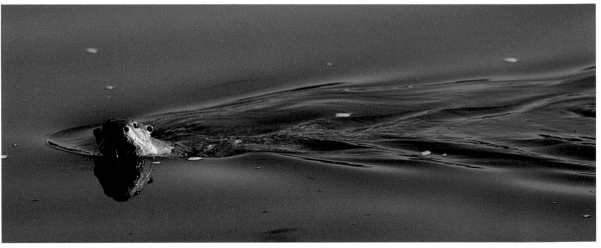

River otter relishes a recent catch. Alan Carey photo.

is usually dark brown, appearing almost black when wet, with a throat area of silver-gray.

These "Lords of the River" are well-equipped to handle their watery environment, with webbed feet, valves that close their nostrils and ears and a thick, insulating layer of fat. They are fast and graceful swimmers, as well as skillful divers capable of staying underwater for as long as five minutes. When there is no reason to hurry, they leisurely paddle with all four feet, steering with their tail. When the mood strikes them, they tuck in their legs and, by undulating their entire bodies in a manner similar to that of humans swimming the butterfly stroke, they can propel themselves through water at up to seven miles per hour. Otters are known to be great wanderers, often covering considerable distances over land, crossing even high ridges on their search for a new river drainage. Surprisingly, river otters, rather than their close relatives the sea otters, are likely to be sighted in the ocean miles from the nearest land, probably on their way to an island.

Lutra canadensis normally has one permanent den, dug into the bank of a stream or lake. Inside lies a cozy nest constructed from a variety of grasses, leaves, reeds and sticks. The nest chamber generally is connected to the outside world by both underwater and above-ground entrances. Although the den is hard to spot, well-worn slides on the surrounding muddy banks betray its presence. In the winter, playful otters may toboggan down long snow-covered slopes on their bellies, with the whole family joining in again and again. Otters are most active during dusk and throughout the night. Those away from home may retire for the day in various resting places like hollow logs, burrows excavated by other animals, or abandoned beaver lodges.

The breeding season for river otters begins in November and extends through early April. Because of delayed implantation, in which a fertilized egg can remain undeveloped for as long as eight months, little ones may not arrive until a year later. The usual two or three offspring are born blind but fully coated by soft fur. At 10 to 12 weeks of age the mother allows them to take their first peek at their surroundings. She calls them to join her in the water and, if that fails, carries the

59

protesting juveniles by the neck to their new environment. The following weeks are spent taking swimming lessons under the watchful eyes of the old female. Young also practice the art of capturing live prey which their mother releases in shallows around the den. Youngsters may be able to look forward to a play-filled eleven years, the maximum lifespan for a river otter in the wild.

It is not surprising that an aquatic carnivore like the otter relies heavily on fish for sustenance, although insects, crustaceans, frogs and the like also are eaten. Sports fishermen traditionally have assumed that this otherwise appealing creature was a major competitor for prized game fish. Research has shown, however, that except in special situations such as along southwestern Montana's trout-rich upper Madison River, slower moving fish like suckers are the most important item in the otter diet.

They have been known to herd a school of fish into the shallow waters of a cove where it is much easier to make a capture. This maneuver may be repeated several times until each has gotten its fill. Sometimes the quarry is eaten while the animal is floating on its back, sea otter style. The carnivores are also at home on land and they often include small mammals, such as mice and an array of plant material in their diet.

Because of their size and aquatic lifestyle, river otters seldom are caught by predators. Intensive trapping severely decimated their numbers before the turn of the century, but state fish and game departments now carefully regulate the harvest of this prized furbearer whose presence may be an indicator of a quality environment. Oregon and Washington account for almost all the Northwest's river otter harvest. Fortunately, unlike in some European countries where lobbying by sports fishermen helped lead to the elimination of river otters, the Northwest still is home to these fun-loving furbearers.

Sea otters spend much of their time resting and feeding among kelp beds. The animals strap themselves seat belt-like to the kelp while napping. Alan Carey photo.

Sea Otter

For ages before white men arrived, Native Americans of the Northwest lived in relative harmony with the sea otter (*Enhydra lutris*). Such tribes as the Coos, Clatsop, Chinook and Quinault had developed moderately efficient sea otter hunting methods. Using spears, harpoons and arrows, they probably were able to keep the otter populations at sufficiently low levels to leave some prized large shellfish for human use.

This delicate balance was disrupted by the 1741 arrival of the Russian ship, *St. Peter,* commanded by Danish explorer Vitus Bering. The Dane had been ordered to explore the unchartered northern seas between Asia and North America, and he carried on board a German naturalist named Georg Wilhelm Steller. During their journey along the Alaskan coast and the Commander-Aleutian islands, Steller saw many sea otters. His account of these mammals was the first written description to be returned to the western world, and was used by Linnaeus when he classified the species.

Unfortunately, the *St. Peter* suffered numerous hardships and eventually was wrecked on the shores of what later became known as Bering Island. This bleak, treeless expanse of sand and rock off the east Siberian coast claimed the lives of Bering and many of his crew during the long winter. Some, including Steller, survived the ordeal and managed, during the next summer, to build a crude vessel from the remains of the *St. Peter.* The sailors thought so highly of the several hundred sea otter pelts they had accumulated that they risked packing them on the leaky tub for the trip back to the Siberian coast.

The risk proved justified. Upon reaching Russia in the fall of 1742, the few survivors sold their *bobri morski,* or "sea beaver" pelts, for a tidy sum. They boasted to their countrymen about the abundance of these animals inhabiting the shallow waters along the Aleutians and North American coast. The news sparked a stampede among the *promyshlenniks,* the Russian counterparts to the later freelance, American mountain men. Within 10 years hordes of small, unstable Russian boats plied the waters of the far Northwest in a marine

Drawn from Nature by J.W Audubon

Sea Otter

Lith. Printed & Colᵈ by J.T Bowen, Philadᵃ

Sea otter, Audubon lithograph. Audubon, Quadrupeds. *Photo from Special Collections Division, University of Washington Libraries.*

search for "soft gold." Many did not return, but some of those that did carried pelt cargoes worth the equivalent of millions of dollars.

Few others knew of these furry riches until the famous English explorer, Captain James Cook, sailed into Northwest waters in search of the elusive Northwest Passage. In 1778 Cook's two ships entered what was later named Nootka Sound on the west coast of Vancouver Island. Before long the sailors were trading the shirts off their backs to the Indians for sea otter skins. Cook recorded, "The fur of these animals is certainly softer and finer than that of any others we know of and therefore, the discovery of this part of the continent of North America, where so valuable an article of commerce may be met with, cannot be a matter of indifference." On the trip home Cook was killed by Hawaiians, but in late 1779, his sailors sold a few worn-out furs to Chinese Mandarins, or nobles, for unheard-of prices. The Oriental demand was apparent, and British and American merchants were eager to provide the goods.

Beginning in 1785, a steady stream of English ships sailed to trade with Indians of the Northwest coast. The Yanks did not get into the act until three years later. In 1789 Spain also joined the fracas, trying to reaffirm her sovereignty over the Northwest by building an outpost at Nootka. Even at this relatively early stage in the trade, the Russians had largely depleted the northern sea otters in the Aleutians and the Gulf of Alaska.

In 1792 when Robert Gray sailed into the lower part of the Columbia River, the more southerly sea otter populations remained largely untapped. Trade inexperienced natives willingly bartered four otter skins for a simple sheet of copper. Just eight days of trading yielded 150 otter, 300 beaver and several hundred other furs.

During the next few years, Russian and Spanish exploitation of Oregon Country sea otters was inconsequential, but British and American competition was fierce. Between 1795 and 1804, 59 British and American ships vied for sea otter pelts along the Northwest coast. Incredible as it seems, Lewis and Clark, after making a journey of 3,000 miles across the wilderness to the mouth of the Columbia, found a declining sea otter population and discriminating Indians who demanded specific trade goods for their pelts. In November 1805

near the mouth of the Columbia, Clark attempted to trade for two otter skins offered by several Clatsop Indians. He noted in his journal, ". . . they asked . . . Such high prices that we were unable to purchase them . . . mearly to try the Indian who had one of those Skins, I offered him my Watch, hankerchief, a bunch of red beads and a dollar of the American coin, all which he refused . . ." Instead, the Indians wanted blue beads, which they were accustomed to getting from the trading ships. In fact, in that same month, the brig, *Lydia*, out of New York, entered the Columbia to trade, unaware that Lewis and Clark were camped nearby at Fort Clatsop hoping for a sea voyage back home for some of their men and scientific specimens.

The onslaught continued, and by the early 1850s Northern Railroad Survey naturalists noted Indians were receiving $30 to $40 in goods for each of the few otters they could find between the Quinault River mouth and Cape Flattery. By that time, Hudson's Bay officials reported almost all their skins came from north of the 50th parallel. They believed otters from the Oregon and Washington coasts had migrated to waters off Japan and Russia.

By at least the late 19th century, the already largely depleted sea otter population along the Washington coast had to contend with one final technological assault—hunters armed with high-powered rifles perched atop seashore derricks. Expert marksmen shooting from these 20- to 30-foot-tall tripods, some equipped with two wheels for ease of movement, may, as at least one author has suggested, been the death knell for the area's sea otter population.

At the beginning of the 20th century, Northwestern sea otter populations that once numbered about 150,000 were reduced to an estimated 1,000 or 2,000 individuals along the northern portion of their former range, and a few hangers-on off California's coast. Between 400,000 and 800,000 pelts are estimated to have been harvested in the Pacific Northwest during the 170 years that followed Bering's discovery. It was a renewable resource that simply could not renew itself fast enough.

Right: September 18, 1897, sea otter stands near Copalis, Washington (about eight miles north of Gray's Harbor). Stanley Jewett, Oregon Historical Society photo, neg. no. 39911.

At last, in 1911, an international agreement known as the Fur Seal Treaty was signed by Britain, Japan, Russia and the United States to protect remaining sea otters. Hunters during the preceding year had caught fewer than a dozen animals throughout the entire hunting grounds. In 1913 California and Alaska passed laws protecting sea otters within their three-mile jurisdictional limits.

For Washington and Oregon the laws came too late. The last native sea otters were thought to have been killed in 1906 at Otter Rock in Oregon, and 1910 at Willapa Bay in Washington. Sea otters did not naturally recolonize these two states' coasts but, of all things, a proposed nuclear bomb test at a wildlife refuge in the Aleutian Islands offered some hope. Amchitka Island was ground zero for the November 1971 test, and Alaskan and federal officials decided it would be wise to remove the endangered sea otters. In 1969, 29 otters were trapped and released near Washington's Point Grenville, known to be preferred habitat. During the following three years, 123 additional sea otters were released off the Washington and Oregon coasts.

The success of the restocking efforts was limited along the Oregon coast with but one sighting in 1981. The story was brighter for otters set free within their historic range in Washington's coastal waters. According to Karl W. Kenyon, retired U.S. Fish and Wildlife Service biologist, 52 otters were counted off the coast in 1983, and an estimated 60 to 70 live there today. Their center of concentration appears to be around Cape Alava and the Bodelteh Islands on the Olympic Peninsula.

Washingtonians lucky enough to observe this only truly ocean-going member of the weasel family would notice it is almost twice the size of its closest relative, the river otter. Sea otters measure up to 71 inches in length, though 50 to 60 inches is more typical. Weighing in at 35 to 90 pounds, they are the smallest of all marine mammals.

The forepaws are stubby and nearly useless for swimming. The "hands" can perform other important tasks, however, such as manipulating food, digging in sand for clams and slamming hard-shelled food items against a rock positioned on the otter's chest. The hind feet are large and webbed, resembling flippers, and are used for aquatic propulsion. When the creature is either resting or eating, it commonly lies on its back, swimming leisurely. However, when speed is required, it flips over to its belly and can, in this position, move about five miles per hour.

Sea otters are covered by an unusually durable, soft and thick fur, considered by many to be the finest in the world. This view was enthusiastically shared by Meriwether Lewis, who entered the following in his journal in February of 1806, "... it is the richest and I think most delicious fur in the world at least I cannot form an idea of any more so it is deep thick silkey in the extreem and strong." Color varies from reddish-brown to almost black, with the head and back of the neck somewhat lighter. Often the older animals, and occasionally youngsters, have a silvery-white head, giving them the nickname, "Old Man of the Sea."

Sea otters are the only marine mammals lacking the insulation offered by a thick layer of blubber. Instead, warmth is provided by air trapped within the fine, dense underfur. This keeps the animals warm and also increases bouyancy. The otter appears to be obsessed with keeping its coat immaculate by constant grooming, but in reality if it were to neglect this task it would soon become chilled and eventually die.

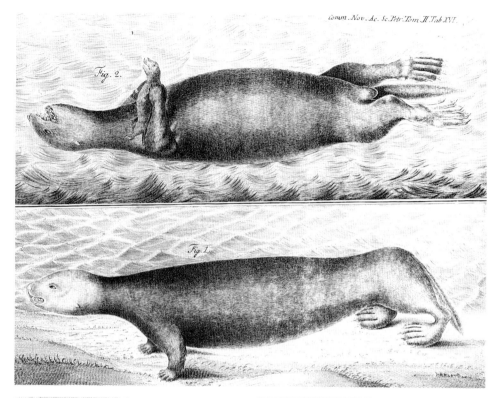

Sea otters, two early depictions. Above right: From Georg Wilhem Steller, "De bestiis marinis," Novi commentarii Academiae Scientiarum imperialis petropolitanae, *(Leningrad: I Akademii Nauk, 1751), pp. 289-293. Photo from Special Collections Division, University of Washington Libraries   Right: From Cook's Voyages, plate 43. Oregon Historical Society photo, neg. no. 3576.*

Mother sea otter carries young on her belly. Alan Carey photo.

Sea otters are highly aquatic and generally take to the shore only to wait out a severe storm or, rarely, to rest at night. Even mating takes place in the sea and can occur at any time of the year. Some 10 to 12 months later the single offspring is born, also usually at sea. The young one resembles a woolly teddy bear with appealing dark eyes. A female sea otter is a devoted mother, often cuddling the little one with her front paws while it nurses, naps or plays securely on her chest.

Sea otters inhabit shallow coastal waters and rarely venture more than a mile or two from shore. They prefer reefs exposed to the ocean, and rocky waters with kelp beds, where their favorite food items, the sea urchins, live. These carnivores usually dive to depths between five and 180 feet for their bottom-dwelling dinner, although one individual was caught in a crab ring 318 feet below the surface in Alaskan waters.

The sea otter has one of the highest food requirements for any animal in relationship to its size. The daily food consumption averages 25 percent of the body weight for a medium-size adult and may be as high as 35 percent for a growing youngster. This means that each individual eats between 15 and 20 pounds of fresh food daily, an incredible

three tons per year! When food is abundant they feast on such delicacies as sea urchins, crabs, clams and abalones if available. They also will eat fish, snails, mussels, squid, octopi, barnacles and even starfish after the choice items have been depleted.

Today sea otter populations along the Northwest coast remain small, and sea otter spotters are only guaranteed viewing if their expedition takes them to the Seattle Aquarium or Tacoma's Point Defiance Zoo and Aquarium. Farther south along the coast, numbers and range of these remarkable sea mammals have expanded to the dismay of shellfishermen. Remember, sea otters have voracious appetites and seek out the same shellfish as do commercial fishermen. Concerned about increased competition from these animals, one of which can consume 80 clams a day or comparable quantities of crab and abalone, area fishermen have organized the Santa Barbara based Save Our Shellfish (S.O.S.) citizen group. They are countered by an equally vocal Friends of the Sea Otter organization based farther north at Carmel. In the 1980s such a shellfishery-sea otter face-off in the Northwest seems improbable.

Cougar

Theodore Roosevelt called the cougar "a big horse-killing cat, destroyer of the deer and lord of stealthy murder with a heart craven and cruel." With indictments like these, it is easy to see why the cougar suffered years of persecution. At the turn of the century, accepted thinking in Northwest state capitals was elimination of this predator through bounties.

Attitudes, of course, change with time, and the legal history of the cougar in Washington serves as an example. From 1905 through 1960 the cat carried a substantial price on its head, increasing from $25 to $75 during that period. Gradually it was realized that the bounty had not eliminated the cougar, and that the animal was not a serious predator on livestock. From 1961 to 1965 the bounty was discontinued, but the cougar still was classified as a nonprotected predator. Once the animal's true resource value was known, it finally was listed as a game mammal in 1966. Today a non-resident hunter must buy $250 worth of licenses ($20.50 for a resident) for the privilege to hunt *Felis concolor*. If the hunter hires a guide with hounds, he may inject at least another $1,000 into the local economy.

Research that helped dispel old myths was one of the main reasons public attitudes about the cougar changed. Probably the greatest contribution came from a long-term, now classic study conducted in the 1960s and '70s by University of Idaho wildlife biologist Maurice G. Hornocker. He and his colleagues worked in the rugged Big Creek drainage along the Middle Fork of the Salmon River in central Idaho, now within the River of No Return Wilderness. Major goals of that landmark study were to learn more about cougar social behavior and predatory habits. Much of the following life history was derived from that work.

If a cougar population is not severely disturbed by humans, a self-regulating mechanism based on territoriality keeps numbers fairly stable. A large area of suitable habitat is divided into home ranges. Those of males' generally do not overlap one another, but they may contain parts of the territories of several females. Adult animals that have secured a domain often are referred to as residents. Young adults, usually transients, are relatively free to move through the established home ranges in their search for an unoccupied area. In such a stable society, fights between males are rare. Each creature evidently avoids direct contact with others, except when a female is ready to mate. Mutual avoidance is maintained by both visual and olfactory clues, which includes scrapes, small piles of dirt or needles scraped up by the resident cat and often marked with its urine. Whenever a wandering cougar smells one of these mounds, it knows it has entered another's turf.

When hunted the cougar's normal organization of home ranges is disturbed by the removal of territorial animals. One consequence may be more fighting between males trying to claim a newly vacated home range. Densities are often higher in this situation than in unhunted populations, possibly because dominant individuals do not live as long and more transients stick around hoping to claim their own territory.

Cougars have no fixed breeding season, rather each female comes in heat on her own cycle once every two years. Whenever this occurs, the generally solitary mammals howl to advertise their desire for company. The resident male joins any such female residing within his domain for one to two days. After a three-month gestation period a litter of two to three spotted young arrives. They spend their first weeks in a den lined with vegetation and hidden inside a cave, under a rock pile or in a brush thicket. The female leaves her young in a sheltered place for up to two days while securing the next meal. At first she drags the carcass back to the kittens, but soon they are capable of following her to the kill site. Gradually she teaches the youngsters the predatory skills necessary for survival, and the young become independent shortly before turning two.

This "lithe and splendid beasthood," as Ernest Thompson Seton described it, measures from five to eight feet from nose to the end of its long, black-tipped tail, and weighs between 130 and 200 pounds. Color varies from reddish-brown to slate

Cougars, also called mountain lions, are surprisingly widespread in the Northwest. Alan Carey photo.

gray, although tawny is most common. The coat is dominated by long guard hairs, which make it coarse and of little commercial value. Cougars have small, rounded ears and large, greenish eyes. Their paws are strong with retractable claws. Although they seldom ascend a tree unless chased by barking hounds, they are good climbers.

The big cat can be found throughout much of the Northwest. They may inhabit areas with dense cover in mountainous forests, but prefer rocky, inaccessible country along rugged, sparsely wooded canyons. Earlier it was believed that undisturbed wilderness areas were key to their survival, but it is now thought they can tolerate man's presence. Their major requirements are an abundance of

suitable prey and absence of persecution. These shy, nocturnal creatures are seldom seen, even when living in close proximity with humans.

Cougars have long been known to prey on big game animals, but Hornocker's studies indicate that severe damage to the herds does not necessarily result. It is true that various large mammals, especially deer and elk, provide most sustenance in winter when a full-grown animal will kill the equivalent of one deer about every fortnight. In most cases, however, this culling is believed to be beneficial to the herds because it keeps the animals on the move. This decreases the risk of overgrazing sections of winter range and removes sick and weak animals from the herd.

The element of surprise gained by painstakingly creeping within a few feet of their prey before attacking is essential in making a successful kill of a large ungulate. Any sizable prey is dragged into cover, where the cougar eats its fill. Leftovers, which may be consumed over a three-week period, are covered with snow or debris and guarded against scavengers. Contrary to their Hollywood image, they do not scream like a human in agony to momentarily paralyze their victim. Lone animals are essentially silent, and a mother and her kittens communicate with birdlike whistles when hunting together.

In summer, smaller animals, including snowshoe hares, marmots, porcupines, birds and even grasshoppers are pursued. When nothing else is available, cougars will eat carrion. Ranchers traditionally considered the big cats a threat to livestock. In the Northwest attacks on domestic animals are rare. Thousands of sheep and probably a couple thousand cougars live together in Idaho, but only one or two sheep predation incidents are reported annually. In Montana, livestock owners seldom have problems with cougars, and the few destructive individuals are easily removed by hunters with dogs. The irony is that an occasional barnyard mutt that forgets its guard duty and remains silent at the approach of a hungry cougar, may end up being the pursuee, rather than the pursuer.

Bobcat and Lynx

For years, European furriers scoffed at the American bobcat and lynx. Their fur was considered brittle, thin and uninteresting, and it was mostly used in cheaper coats or for trim. Elite fashion houses preferred the much more romantic and rare leopard, jaguar or ocelot with which to drape the shoulders of the world's well-heeled.

Fur buyers had a sudden change of heart in 1976 when international trade in many exotic cats was curtailed under the provisions of the Convention on International Trade in Endangered Species of Wild Fauna and Flora (the CITES treaty). Trade in bobcat and lynx pelts still was allowed, and those wanting spots on their fur coats extolled the animals' virtues. An American trapper who received only $10 to $15 for a bobcat skin in 1971 found it brought more than $300 in 1979. An excellent lynx pelt brought up to $800 by the mid-1980s. Considering that many of the best coats are made from only the cat's belly fur, it is easy to see why such garments command astronomical prices.

The spotted pattern so highly valued by the Jet Set is of even greater value to the intact cat. The light and dark pattern helps these animals blend almost perfectly into their surroundings when the sun's rays filter through a dark forest canopy. Most bobcats have coats with a gray or tan tint, but those living west of the Cascades have a distinct reddish hue, which gave them the old name, "red cat." This 15- to 35-pound animal has a stubby bobbed tail. Hair tufts on ears are small or may be lacking. Although the lynx weighs 15 to 30 pounds and measures three feet in length, it appears much larger because of its longer legs and thicker fur. Lynx coats are tawny or pale gray, speckled with blackish hairs. Spots are indistinct except on the belly. The short tail is encircled with black at the tip and the black ear tufts are more prominent.

Bobcats are the most numerous of all wild cats in the United States south of the Canadian border. They occupy favorable habitats throughout the Northwest. The cats have fared better than other large felines because they are more tolerant of human presence. Some individuals even have set up housekeeping under abandoned houses near

Bobcat, Audubon lithograph. Audubon, Quadrupeds. *Photo from Special Collections Division, University of Washington Libraries.*

large cities. Since the bobcat eats any small mammals it encounters, the clearing of virgin forests by logging or for agricultural purposes actually benefits this carnivore by increasing the abundance of rodent prey species.

The lynx has suffered a loss of habitat due to the same activities that enhanced the survival of its cousin. This species prefers the dense, pristine, mature coniferous forests of the backcountry, and retreats from the advance of human settlement. Today its range covers only the northern parts of Washington (seldom west of the Cascades), northern Idaho and the northwestern corner of Montana.

Individuals of both species are solitary hunters of the night. Both hearing and eyesight are extremely well developed. They rely mainly on the two senses in their efforts to locate prey, unlike many other carnivores that track their victims by scent. The long whiskers adorning their faces are exceptionally sensitive to touch and help the mammals maneuver in the dark.

Cats sometimes lie in ambush along rabbit trails, patiently waiting for an unsuspecting victim to hop by. Another typical hunting technique is the well-known feline stalk where the cat creeps slowly, belly pressed against the ground, toward its goal. It uses every available shrub and rock as cover to move within a short distance of the prey, then pounces with amazing speed and accuracy. Common species caught by the bobcat are various rabbits and hares, especially the brush rabbit along the coast. Mice, voles, mountain beavers and ground-nesting birds also are eaten.

Opposite page: Adult lynx at sunrise, Flathead River, western Montana. Alan Carey photo Above: Bobcats are the most numerous of all Northwest wild cats. Alan Carey photo Above right: A litter of two kittens suggests only an average population of snowshoe hares, the lynx's primary prey. Alan Carey photo Right: Bobcat kittens at play. Alan Carey photo.

The lynx specializes in one major food item—the snowshoe hare. Its oversized feet resemble those of its prey, enabling the cat to move easily through deep, soft snow. Because of heavy dependence on a species whose numbers fluctuate widely, lynx populations undergo similar drastic changes. When at their peak, snowshoe hare numbers may be 20 times higher than during a low, which may have occurred only five years earlier. A wooded area that supported four lynx in good years may sustain one or none during lean times. A year or two following a peak in hare numbers, the lynx population reaches its highest density. During these times the normally wary cats may spread from their forested retreats to more open areas, or even into towns.

Each lynx is believed to require about 200 snowshoe hares per year, but they supplement their diets with other small animals. Both the bobcat and lynx sometimes kill larger mammals including deer, especially fawns. The cats often cover leftovers from such sizeable catches by scratching loose sticks and soil upon the carcass.

These cats are generally silent, but in spring the loud yowling of the males announces that the breeding season is underway. Dens usually are within a cave, in a crevice among rocks or in a hollow tree. The female bobcat delivers an average of three or four kittens; the lynx may have four offspring during a season of abundant snowshoe hares, or just one when they are scarce. Soon after their birth, the mother cat serves them small live mammals. This allows kittens to practice the stalking and pouncing behavior so vital for their future survival.

Many of these cats, especially the young, probably die from starvation rather than predation. Enemies are cougars and rare wolves. Occasionally a great-horned owl might include a careless kitten in its menu. Humans have a history of ambivalence toward these wild cats. In the Northwest they have been legally bountied as a nuisance predator, classified as furbearers, ignored or protected. Today the bobcat is considered a furbearer in all four states with an annual harvest of 5,000 to 6,000, half from Oregon. Only about 100 of the much rarer lynx are trapped annually in the Northwest, almost all from the wilds of western Montana.

69

Sea Lions and Seals

California and Steller's sea lions at Simpson's Reef south of Coos Bay, Oregon. James T. Harvey photo.

Steller's Sea Lion and California Sea Lion

Late spring and early summer are hectic times in the lives of sea lions. In early May the giant male Steller's sea lions (*Eumetopias jubatus*) return to their traditional breeding grounds on rocky offshore islands. A mainland rookery, such as Sea Lion Caves 12 miles north of Florence, Oregon also may be used. Some 3,000 of these marine mammals breed off the Oregon Coast and a few hundred more frequent the coastal and inland waters of Washington, where some can be seen on the Jagged Islands and the outermost Flattery Rocks.

The first few weeks following the males' arrival are spent feuding over choice territories, since female sea lions seem to be more attracted to a specific turf than a macho male. By the late-May females' arrival boundary disputes have been settled, and territory is maintained by threat displays and near-constant roaring. Although males are capable of breeding around their fourth birthday, they lack the competitiveness and size necessary for claiming turf for another five years. Territorial males, or beachmasters, often fast the entire 60-day breeding season, because leaving their ground and harem of 10 to 20 females would be an open invitation for a usurper.

Most Steller's pups are born in late May to early June, after an 11½-month pregnancy that includes a period of delayed implantation. To achieve the same results as the proverbial slap on a newborn's bottom, the mother sea lion lifts her pup a few feet off the ground and lets it fall on the rocks. One devoted female was observed repeating the act 52 times before being satisfied with her offspring's response. The mother recognizes her pup by smell and voice, but the pup initially has difficulty locating the correct female. If the youngster tries to solicit food and sympathy from another's mother it may end up being hurled through the air.

Dark brown newborns weigh about 40 pounds and average 3½ feet in length. They grow rapidly, doubling their weight in seven weeks. Initially, pups are afraid of water and play only in the shallows where flippers touch bottom. Mother has to encourage them to enter the sea when they are four weeks old, but as soon as they gain confidence they obviously enjoy frolicking in deeper water.

California sea lions (*Zalophus californianus*) breed mainly on the coastal islands of California with somewhat more loosely assembled harems. Otherwise their breeding and pupping closely resemble that of the northern sea lions.

Males of both species head north after the breeding season, while females with young remain close to the rookeries. Steller's males migrate to British Columbia and Alaska leaving few, if any, in the coastal waters of Oregon or Washington after the end of October. About 5,000 of their southern relatives travel more than 600 miles north from their California breeding grounds and mingle with the northern species off coastal British Columbia. Some of the California travelers also winter along the Washington and Oregon coasts and inland Washington waters. The eye-catching cavorting of these migrant males easily is observed in many Northwestern bays and rivers. According to Robin Brown, marine biologist with the Oregon Department of Fish and Wildlife, California sea lions may occur along the Oregon Coast from September through May. Their numbers peak in December and January when as many as 150 gather in Newport's Yaquina Bay.

In Washington these mammals lately have been moving farther south into Puget Sound from their traditional wintering grounds among the San Juan Islands. Around 200 were counted in the Nisqually area in April of 1984. This expansion possibly is related to increases in the Sound's hake or squid populations. Sea lions have learned to use man-made objects in addition to beaches for sunning. Individuals can be seen basking in the sun atop a swimming raft off Redondo Beach near Federal Way, Washington, and others off Everett readily climb aboard an abandoned off-shore barge provided courtesy of the local Friends of the

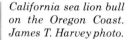

California sea lion bull on the Oregon Coast. James T. Harvey photo.

Steller's and California sea lions at Sea Lion Caves north of Florence, Oregon. Sea Lion Caves photo.

Sea Lion group. Most of these seasonal visitors return to California shores by May, although some have made the Northwest their permanent home. One old individual locals call Oscar spends the entire year near Olympia, Washington where it occasionally dines on salmon migrating to Capitol Lake.

The Northwest's two sea lion species can be distinguished by differences in size and coloration. The Steller's sea lion, named after German naturalist, Georg Wilhelm Steller, is the world's largest. Males weigh up to a ton (as much as a bull bison) and measure 13 feet in length. Their forequarters are massive, and the muscular neck is covered with a long, coarse mane. Their female counterparts are slimmer and smaller, seldom exceeding eight feet long and 600 pounds. Pelage of adult individuals varies from yellowish-brown to tan. California sea lions are more slender, with males roughly the size of northern sea lion females. Coloration of adults is normally darker than their northern cousins, appearing black when wet.

Both the Steller's and California sea lions are opportunistic carnivores. Some scientists believe they feed mainly on marine invertebrates, such as mollusks, squids, octopi and worms, although many fishermen blame sea lions for declines in the numbers of commercially valuable fish. A few may learn to raid gill nets, and salmon or steelhead may be eaten occasionally, but most prefer hake, rockfish, herring and other slower moving, non-commercial species. Two Steller's sea lions

killed near the mouth of the Columbia River at the height of a salmon run had filled their stomachs entirely with lampreys! California sea lions feed at night, eating small items at the bottom, but carrying larger ones to the surface where they are ripped into pieces before swallowing. Steller's sea lions are known to ingest stones that may be picked up accidentally when they feed on bottom-dwelling organisms. Some think the stones serve a specific purpose, perhaps keeping the animals from feeling hunger pangs, especially during the 60-day fast endured by beachmasters, or helping grind large pieces of food.

Sealers of the 19th century, who figured it took the blubber of three to four adult male sea lions to make a barrel of oil, badly overhunted the mammals. Today sealers are gone, but killer whales and white sharks still prey on sea lions. Irate fishermen also may be an important source of mortality, as one marine biologist noted that 30 percent of all sea lion carcasses collected along the Oregon Coast showed evidence of having been shot.

These mammals are highly gregarious, spending most of their time in small groups at sea, but seldom venturing more than 15 miles from land. When weather is sunny and calm, groups may haul out on a rocky beach. They generally are playful, and California sea lions often can be seen leaping out of the water while swimming at speeds up to 17 miles per hour. Their friendly nature has made the California sea lion the favorite performer in aquariums and as circus "seals." Military technicians have taken advantage of their ability to stay submerged for 20 minutes and dive to depths of 650 feet by training them to aid in searching for lost missiles and torpedoes. Presently two even are being trained in California to help rescue drowning swimmers.

Harbor Seal

Above: Harbor seal pup at Mark O. Hatfield Marine Science Center in Newport, Oregon. As a participant in the Center's rehabilitation program, this abandoned pup was kept about two months. James T. Harvey photo.

Right: A word of warning to well-intentioned beachcombers. Oregon State University Extension Service.

**Seal pups rest on shore
Do not disturb them !
It's the law.**
Report animals in distress
to the Oregon State Police

**Oregon State University Extension Service
Sea Grant Marine Advisory Program
SG 69 July 1981**

Beachcombing along a deserted stretch of Northwest Coast always offers the hope of discovering some small bauble, perhaps a colorful shell or a glass Japanese fishing float. Occasionally, the sea delivers a living treasure to the beach, sometimes in the form of a harbor seal pup. The big brown eyes seem to plead, "Take me home, I'm lost." Wise shoreline visitors know, however, that young seals may become temporarily separated from their mothers in the near-shore surf, or they may have been left there to rest while the mother forages. Less than half the youngsters "rescued" by well-meaning humans truly are abandoned by the parent.

Generally, young seals, or sea lions, should be left alone for at least 24 hours after they have been found. If they have not returned to the sea by then, the concerned citizen should call the Oregon State Police or Washington State Patrol. Both agencies then contact representatives of the Northwest Marine Mammal Stranding Network, a group of experienced scientists and interested laymen who volunteer their time to help stranded marine mammals. If the animal dies, they see that useful scientific information is obtained from the remains, and they help coordinate proper disposal procedures.

The harbor seal (*Phoca vitulina*) is the Northwest's most abundant sea mammal, and is the only seal that breeds along the coastal waters of Oregon and Washington. Each state has an estimated 4,000 harbor seals, with as many as 1,500 of them gathering at the mouth and in the lower reach of the Columbia during salmon runs. This is where Lewis and Clark saw them in October 1805. Initially Clark misidentified them as sea otters, but corrected himself while wintering at Fort Clatsop after having seen real sea otters. In February 1806 he recorded in his journal, "Those animals which I took to be the sea otter from the Great Falls of the Columbia [Celilo Falls] to the mouth, proves to be the Phosis or Seal. . ."

Both sexes of harbor seal are of nearly equal size, about four to five feet long and weigh around 250 pounds. Their eyes are large and external ears

are absent. Front flippers are small while backward-pointing large hind flippers create the propulsion necessary for swimming. Their short-haired coat is highly variable in color, ranging from a silvery gray to dark brown, with irregular spots of brown or gray.

These stocky mammals may be seen resting in large groups on tidal flats, exposed reefs or rocks during low tide. They are fairly antisocial, however, and such groups appear to be the result of limited hauling out grounds rather than a need for company. As the tide comes in, individuals slide back into the water and go their separate ways, each foraging in solitude.

Although major food sources are bottom-dwelling fish, crabs and herring, harbor seals also dine on salmon and can be destructive in their efforts to relieve gill nets of their catch. For years seals were persecuted by fishermen using traps, guns and even dynamite. Both Washington and Oregon ran official harbor seal eradication programs to lessen their impact on fisheries. Since the 1972 passage of the Marine Mammal Protection Act numbers have risen steadily.

Before legally protected, the animals were harassed constantly in the sheltered estuaries, their preferred reproductive habitat. Since 1972 seals have been able to rear their young in peace which has meant increased production. In the late 1970s, 400 seals made their home in Oregon's Umpqua River estuary. By 1984 around 1,000 harbor seals were counted in the same waters. Similar increases in Washington's Willapa Bay and Gray's Harbor, as well as the lower Columbia have led to more conflicts between fishermen and seals.

Antisocial by nature, harbor seals prefer to forage in solitude. Thomas L. Spaulding photo.

Human-caused fatalities remain a population factor. It is estimated that 200 to 300 seals are shot annually by commercial salmon and sturgeon fishermen along the Columbia River alone. Commercial operators may obtain a Certificate of Inclusion which allows them to take netted animals incidental to catch, and also may kill seals if they threaten fishing gear or the catch. All such seal deaths are to be reported to the National Marine Fisheries Service although many are not.

Predators of seals, including killer whales and sharks, evidently are not keeping the animal's numbers in check. Because of fishing pressure controlled harvests of seals at such places as Yaquina Bay and Gold Beach, Oregon might be appropriate in the future, but seal skins are nearly worthless and the oil market is poor.

Above: Harbor seals at Tacoma's Point Defiance Zoo and Aquarium demonstrate how hind flippers are used for propulsion while front flippers serve as rudders. James T. Harvey photo.

Right: Elephant seal found in Strait of Juan de Fuca. These large seals may be more common than previously thought in Washington's inland waters. Boaters often mistake these up to 20-feet-long and 8,000-pound sea mammals for floating logs and steer clear. Ken Balcomb, III photo.

Whales, Dolphins and Porpoises

Spyhopping Puget Sound killer whale. Ken Balcomb, III photo.

Gray Whale

Roadside signs announcing "Whaler's Rest," "The Whaler Motel" and "The Whale's Tail Restaurant" along U.S. Highways 101 and 109 leave little doubt that Northwest residents are friends of the gray whale. Whale watching is more than just a local custom, it is a source of livelihood for some industrious citizens. As salmon fishing declined over the last several years, charter boat owners found they could pick up some of the slack by taking out as many as a thousand eager whale watchers a season. Two- to three-hour whale watching excursions operate out of several ports, including Coos Bay, Newport, Depoe Bay and Lincoln City in Oregon and Westport, Washington.

Knowledgeable whale watchers delight in impressing visitors with the fact that passing gray whales are undertaking the longest annual migration known for any mammal. From their summer feeding grounds in the icy waters of the Chuckhi, Beaufort and Bering seas above the Arctic Circle, they follow the shorelines south to their calving grounds along Baja California. When on the move, the animals cruise at 4 to 5 miles per hour, and cover the 5,000- to 6,000-mile one-way trip in about 100 days.

Especially on their northward migration gray whales (*Eschrictius robustus*) travel close to the Northwest Coast, often passing within two to three hundred yards of many headlands. The mammals easily can be observed from such promontories as Cape Flattery, Point Grenville and Cape Disappointment in Washington, and Tillamook Head, Cape Perpetua and Cape Blanco in Oregon.

Employees with Oregon State University's Extension Sea Grant Program at internationally known Hatfield Marine Science Center in Newport offer workshops, lectures and films about whales for the public. Starting in 1984 volunteers who attended one-day training sessions on gray whales have been stationed at numerous strategic lookouts along the Oregon Coast during the height of the spring migration. Whale watchers interested in learning more about the grays thus may turn to nearby experts for factual information. Washingtonians desiring an in-depth knowledge may enroll in a week-long Whale School at Friday Harbor's Whale Museum. This summer workshop, which includes field work and a comprehensive lecture series, may be taken for college credit through Western Washington University.

Marine biologists think grays' winter migration to warm southern waters may be linked to calving, since young are born with little insulative blubber. Not surprisingly, pregnant females are the first to head south, usually in October. Soon adult males and other females follow suit, while adolescents bring up the tail end. The first whales appear off Washington's coast in late October, and by mid-December most migrants have passed. Oregonians can start looking in earnest for the steamy spouts in November, and the majority of grays are gone by early February.

Some seniors and juniors evidently do not make the complete migration, opting instead to linger behind and explore the bays and coves en route. Adult females that have already visited Baja, as well as still-attentive suitors, commonly start heading north in February, even before all youngsters have reached wintering lagoons. Still-southward-bound juvenile groups encountering these adults reverse direction and join those on their return to the Arctic waters. Cows that have given birth to their single calves do not leave Baja until May. Thus the northward spring migration, the one that most interests whale watchers, extends from early March through June in Oregon, while

Left: Gray whale in San Ignacio Lagoon, Baja California, Mexico. Scientists study movement of grays by attaching radio tags to the giant mammals' backs. James T. Harvey photo.

77

Left: Off the Oregon Coast a mother gray approaches a spyhop as her calf surfaces. Eva Cooley photo Above: In a Baja lagoon a female gray whale gives birth to a 2000-pound baby. A head-first delivery is rare for baleen whales. Ken Balcomb, III photo.

whales reach, and leave, Washington waters a few weeks later. Exceptions include the small population of grays, perhaps those too young or too old to make the entire trip, who evidently summer in the Strait of Juan de Fuca as well as farther north off the west coast of Vancouver Island.

Gray whales make their odyssey alone or in groups of as many as 16. A traveling whale usually makes a short, four-minute shallow dive, then surfaces and blows a few times. Several such dives often are followed by sounding, which is a longer (up to 15 minutes) and deeper dive. Due to steepness of the descent, tail flukes may emerge from the water, as if the animal were standing on its head. Migrating whales spyhop, possibly to determine their location relative to the coastline. Sometimes grays breach, with the animal lifting half or more of its body above water, and then rotating to its back or side before falling with an impressive splash. An individual often repeats the act, and others may join in. Explanations for this striking behavior range from communication to an attempt

at getting rid of whale lice, or the animals may just be having fun.

Gray whales are medium-sized representatives of the baleen whale group. Like their relatives they are toothless, relying instead on strips of whalebone, or baleen, hanging from the roof of the mouth to strain food from the water. They may be 35 to 50 feet long and weigh 20 to 40 tons, with females, for a change, considerably heavier than males. They lack a dorsal fin, but there is a slight hump, followed by a series of bumps, or knuckles along the dorsal ridge of the tail stock. Gray whales are a basic dark slate gray, heavily mottled with lighter natural pigments and up to hundreds of pounds of barnacles. Many a gray whale also has white scars adorning its flukes and flippers, souvenirs of escapes from killer whales and possibly large sharks.

Most young are born in one of five major Baja California calving sites. Best known is 30-mile-long Scammon's Lagoon, designated a whale refuge by the Mexican government in 1974. Here and

in the other lagoons the female, which has carried her young one for 13 months, delivers a 15-foot-long, 2,000-pound baby. Each day, the mother squirts some 50 gallons of extremely nutritious milk into her offspring's ever-willing mouth. It contains more than 40 percent fat (human milk has 3½ percent) and has about twice as much protein as the milk of most terrestrial mammals. On its rich diet the calf gains up to 70 pounds a day. These large mammals may live 60 years and sometimes continue growing throughout the first half of their lives.

Because they migrate close to shore, gray whales were hunted for centuries by aborigines along the Northwest Coast. Recently, Richard Kool of the British Columbia Provincial Museum has questioned the common belief among anthropologists that the gray was the primary quarry of the Northwest's whale-hunting Indians. He thinks Indian artifacts, stories and early pictures suggest that humpbacks were most hunted. Whichever species dominated, the Makah-Ozette, Quillayute-Hoh and Quinalt-Queets regularly whale hunted, with the activity most important to the Makah. H.H. Webster, Makah Indian Agent, observed in 1865, "What the buffalo is to the Indians on the plain, the whale is to the Makah."

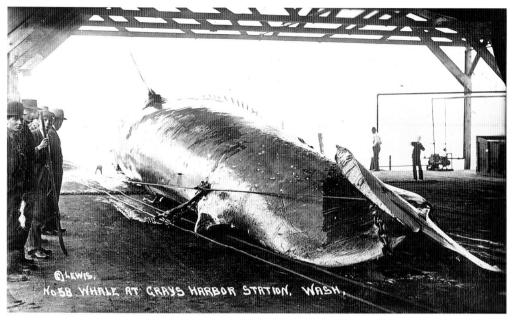

No 58 WHALE AT GRAYS HARBOR STATION, WASH.

Above left: Six species of whale totaling 2,698 individuals were received at the Grays Harbor Station, Washington between 1911 and 1925, when two to four killer ships operated per season within 150 miles of port. Wesley Andrews, Oregon Historical Society photo, neg. no. 29586 Above: The steam whaler, "Aberdeen," was assigned to the Grays Harbor Whaling Station. Still afloat at the onset of World War II, the craft was taken over by the military for use as a Coast Guard patrol boat. Special Collections Division, University of Washington Libraries Left: Makah Indians of Cape Flattery, Washington section a gray whale. The chief marked each section, the allotment of portions depending upon the position of each Indian in the canoe. Asahel Curtis, Washington State Historical Society photo.

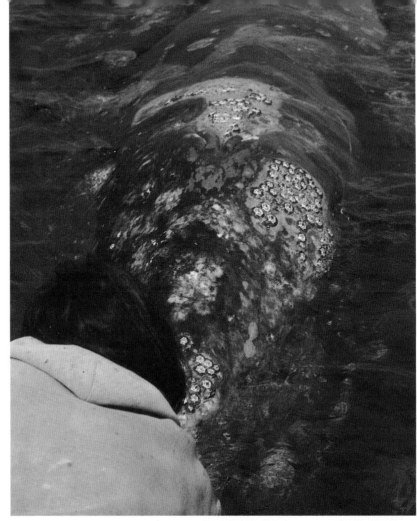

The Makah used an eight-man version of the cedar Nootka canoe for whaling, primarily in May and June. An elk antler or bone harpoon head was plunged into the whale by use of a shaft. Braided bark lines connected the harpoon head to seal skin floats, which helped to prevent the animal from diving effectively, or sinking when it died. Three or four of the huge sea mammals would sustain a Makah village for a year, and excess meat and oil were bartered to inland tribes or, later, to commercial traders. Numerous displays relating to the rich and spiritual Makah whaling tradition can be seen at a tribal museum at Neah Bay on the tip of the Olympic Peninsula.

Intensive American and European whaling for grays in Baja began in the 1840s, but whalers considered the oil inferior and most preferred taking the larger and less dangerous right and bowhead whales in northern areas. Interest in grays was renewed following the 1857 discovery of Scammon's Lagoon by whaling captain Charles M. Scammon. The lucky strike coupled with new hunting methods led to merciless slaughter in all lagoons.

By the late 1800s gray whale population along America's West Coast was so low that commercial harvesting no longer was profitable. After a brief

Far left: "Sea Pirate" out of Newport, Oregon readies for a whale-watching charter. M. Douglas Scott photo.

Left: Grays often provide free transportation for hundreds of pounds of barnacles. James T. Harvey photo.

respite, whaling was resumed in 1914 by modern "floating factory" methods and continued sporadically, despite a 1937 international agreement designed to limit the harvest. Scientists feared grays were nearly extinct that year after none were sighted along the entire migration route. In 1946 the International Whaling Commission granted full protection, although Siberian Eskimos, or their Russian guardians, were allowed to harvest up to 200 grays annually. The remnant population was estimated anywhere from a few hundred to a few thousand animals when finally protected. Grays quickly rebounded with the population now thought to total between 16,000 and 17,000—perhaps equaling the number prior to commercial exploitation.

A 1978 International Whaling Commission decision allows the Soviets to harvest 178 gray whales per year for aboriginal use, and Alaskan Eskimos are permitted to kill several annually. Such a low harvest probably has little impact on the herd. The only major threat to survival would be unusual activities such as offshore drilling that would harm the feeding or breeding grounds. Harassment of whales and calves in the Mexican lagoons by sea-going sightseers has been a problem. Some think salt processing at Scammon's Lagoon has driven gray whales away from this preferred calving ground. Increased protection of these lagoons has become a higher priority with the Mexican government, as well as with the Northwest's loyal contingent of whale watchers.

Killer Whale

Some biologists who study killer whales complain the animal suffers an image problem. The name does not have a friendly ring to it, and some prefer the animal be called by its specific name, orca. Without doubt, killer whales (*Orcinus orca*) once were considered among the most bloodthirsty, merciless man-eaters that ever roamed the oceans. These "wolves of the sea" long have been known to prey on other mammals, and it was naturally assumed that man would likely receive similar treatment. Sources as diverse as Eskimo legends and U.S. Navy diving manuals have served to reinforce the idea. Such attitudes changed rapidly, however, after several killer whales caught alive and kept in aquariums proved to be highly intelligent and quite gentle. Thorough searches of the records also showed that, although attacks on small boats have been recorded, no irrefutable instances of orcas attempting to kill humans are known.

Like many cetaceans, male orcas are considerably larger than females. Mature bulls may attain 30 feet in length and 6 tons or more in weight, while a large cow might measure 21 feet long and tip the scales at 8,000 pounds. Part of the mass is contributed by 5 to 8 inches of insulating blubber. This largest member of the dolphin family is unmistakable due to its size, combined with a striking color pattern. Upper bodies are black and underparts a starkly contrasting white. The white extends onto the flanks where there is a patch behind each eye. The triangular dorsal fin may measure up to six feet tall on older males. Behind the fin lies a grayish, saddle-shaped patch.

Unfounded fear of this sleek, black and white sea mammal was just one example of how little has been known about whale biology. In an effort to solve some of the mysteries, a few Washington scientists formed the Moclips Cetological Society in 1969, since relocated on San Juan Island in the village of Friday Harbor. Now more commonly known as The Whale Museum, this non-profit corporation conducts scientific investigations, operates a Whale School, publishes several periodicals including scholarly *CETUS: The Journal of Whales, Porpoises and Dolphins* and operates the year-round Whale Museum at Friday Harbor. The museum's giant Hotline Map is one of the most popular displays with the 20,000-plus yearly visitors. It locates the latest sightings of killer whales as well as other whales and porpoises in the greater Puget Sound area based on western Washingtonians' reports called in to the Whale Hotline number (1-800-562-8832).

The Whale Museum's most publicized investigation to date undoubtedly is their study of killer whales frequenting Puget Sound. The first task facing Ken Balcomb, III, Moclips president and principal investigator between 1976 and 1983, was to identify individuals. Using photo-identification techniques developed by Dr. Michael Biggs at The Pacific Biological Station at Nanaimo, British Columbia the scientists took and analyzed over 17,000 photos. Eventually 79 residents were identified. Subtle differences in the saddle-shaped patches and in the heights and scarring of the dorsal fins helped to distinguish animals. Such markings are thought to be nearly as reliable as human fingerprints.

Long hours of observation of identified individuals revealed that animals live in three pods, or family groups, designated J, K and L. Each group differs somewhat in its behavior. J-pod does not seem to mind tourists with their clicking cameras, while the larger L-pod spends little time in the busy inland waters around the Sound and, like K-pod, may feed along the western coasts of the Olympic Peninsula and Vancouver Island in winter and spring. Small transient pods, usually groups of fewer than six individuals, also frequent

Killer whale, the "wolf of the sea," prowls Puget Sound. Ken Balcomb, III photo.

On tour at Friday Harbor Whale Museum. Howard Rosenfeld, Whale Museum photo.

Puget Sound. Unlike locals these out-of-towners, which may range as far as Alaska in their sojourns, do not travel set routes and can be found in almost any section of the Sound.

Killer whales are opportunistic feeders, concentrating on whatever is most available. They favor the shallow waters of bays and estuaries that team with animal life. Pods residing within Puget Sound subsist mainly on fish, especially salmon. In addition to fish, transient pods derive their sustenance mostly from other warm-blooded animals such as seals, porpoises, dolphins and even other whales. Some biologists believe this is a result of the resident pods claiming the better fishing grounds, thus leaving the transients to feed on miscellaneous foods in poorer sites.

According to Robin Brown, marine biologist with the Oregon Department of Fish and Wildlife, one such transient group began including the coastal waters of Oregon in its far-reaching travels in the early 1980s. Since then a few orcas have been sighted every April in such places as Tillamook Bay, Coos Bay and the mouth of the Siuslaw River, at Florence. When a small pod publicly seized a couple California sea lions next to the docks in Newport's Yaquina Bay, pier-side spectators had mixed emotions.

There is no doubt that orcas are efficient sea hunters. A Dutch zoologist reported that a 16-foot killer whale was found stuffed with the partial remains of 14 seals and 13 porpoises! Occasionally orcas prey on other whales several times their size, including gray whales. Orcas direct their attacks on the flukes and flippers of the large whales and

Adopt an Orca

Begun in 1976 the killer whale research by Washington's Moclips Cetological Society scientists has evolved into one of the most important studies of its kind. More is known about Puget Sound's resident orcas than almost any other whale community in the world, but much additional research is needed. You can be a part of this exciting and pioneering work.

To more directly involve the public and to help raise research funds, the Society launched its innovative Orca Adoption Program in 1984. Now, for a one-time fee of $20 ($50 for a business) you can adopt a local Puget Sound orca. Individuals are free to select their adoptee from a list of all 70-odd resident orcas, each identified by name (Lulu, Mozart, Sissy, Sonar, etc.) and grouped into males, females and calves.

Adopters receive a frameable, 8½" x 11" certificate including the orca's identification photo, information on its characteristics and social relationships within his or her pod and an annual *Orca Update,* a fall newsletter highlighting the previous summer's field research. Individuals, classrooms, businesses, service organizations and others wanting to adopt an orca can write:

ORCA ADOPTION PROGRAM, The Whale Museum, P.O. Box 945, Friday Harbor, Washington, 98250.

"Olympia," a female member of L-pod and recent adoptee of Northwest Panorama Publishing, Inc. Frederic Felleman photo.

Congenial members of J-pod cavort in Puget waters. Howard Rosenfeld photo.

for some reason commonly only eat flesh from the lower jaws and tongues of their victims. Calves of the large whales also may be taken.

Killer whales are known to emit three distinct types of sounds that can be detected by underwater hydrophones. Clicking noises are used as a sonar to locate prey, while various "screams," or pulsed sounds, evidently help the animals maintain contact when out of each other's sight. Use of whistles, which usually are associated with social interactions, are poorly understood by humans. Such language has distinct regional dialects, and varies between pods residing in the same area. Each individual orca also has vocal characteristics that are special to it, enabling whales to tell each other apart at a distance. Evidently dialects are stable, allowing Skana, a captive orca at the Vancouver Public Aquarium in Canada, to "speak" the same dialect as his pod members after 14 years of separation.

Orcas are among the fastest swimmers in the ocean, sometimes reaching speeds up to 30 miles per hour, although a typical cruising speed is a modest five miles per hour. These mammals are very playful, commonly performing acrobatic somersaults, tail lobs and spyhops. They sometimes synchronize their actions, such as surfacing simultaneously to take a breath. At rest, pod members form a tight group and swim slowly just below the surface.

Killer whales breed any time of the year but sexual activity peaks around early September in Puget Sound, when pod communities come together to cooperatively feed on salmon. The single calf is born following a 16-month gestation period. The newborn measures eight feet long and weighs about 400 pounds. Normally the calf can expect to stay in the same pod for the remainder of its life. Cows may give birth only once a decade, though some have calves at three-year intervals. If the orca reproduction rate is one of the lowest among social animals, so, too, is its mortality rate, presuming no human-caused deaths. A bull may live as long as 50 years while cows are thought occasionally to reach the century mark.

In the 19th century orcas were left in peace because of their poor yield of oil and the fact that their quickness in shallow coastal waters made hunting difficult. The Makah Indians living near Cape Flattery on Washington's coast occasionally took a few because they considered the orca's flesh and blubber superior to that of bigger whales. Norway, Japan and the Soviet Union began harvesting orcas after the larger and more profitable whales became scarce, but the International Whaling Commission now has placed a ban on open-seas whaling for orcas. With the 1976 passage of legislation making live capture of orcas in Washington waters illegal, the killer whale's future in the Northwest seems secure.

Harbor Porpoise

The harbor, or common, porpoise is the smallest and most abundant of all marine cetaceans in Northwest coastal waters. These small porpoises frequent the mixed salt and fresh waters of many bays and harbors and occasionally even ascend the larger rivers to the head of tide. They rarely venture more than 20 miles from shore.

Harbor porpoises (*Phocoena phocoena*) are relatively shy mammals. They seldom play around boats and do not ride bow waves as do many other members of the dolphin family. Even in busy harbors all most people manage to see of them is a dark back with a low, triangular-shaped dorsal fin, revealed as the animal barely surfaces for a breath. Between dives they usually swim close to the surface and rise above the water approximately four times per minute for air. They have quick and quiet blows when breathing, hence their nickname, "puffing pig." After taking a deep breath an individual may dive to depths approaching 44 fathoms.

These mammals are not as gregarious as many of their relatives, and typically spend their time alone or in small gams of two to six. When a large school of herring strays into the coastal waters inhabited by these creatures, as many as 30 to 50 porpoises may gather in an area less than one-half square mile in size, feeding frantically on the abundant prey. When the fish move on, harbor porpoises promptly disperse.

Nondescript, they are slate gray to nearly black above, with light gray to whitish undersides. They may reach an overall length of six feet, but 4½-foot individuals are more common. They average 120 pounds, often with more than 40 percent of the body weight comprised of a thick, insulating layer of fat. Harbor porpoises have a short, beakless snout and jaws containing flattened, spade-like teeth.

Spyhopping Puget Sound killer whale. Ken Balcomb, III photo.

A fleeting glimpse of a dark-gray back and short dorsal fin is about all mariners manage to see of the shy harbor porpoise. Ken Balcomb, III photo.

These marine carnivores feed mainly on fish, including herring, hake and pollock. A few squids and an occasional crustacean may be added to the menu. Because their teeth are practically worthless for chopping food, they swallow their prey whole. A few incidents of overly greedy individuals choking to death on too big a bite, such as a small shark, have been recorded.

Porpoises use a highly evolved sonar system to locate their prey. They emit various clicking sounds that bounce off a fish or other animal, leading the predator to its dinner. This echolocation ability is so finely tuned that the harbor porpoise can detect and avoid metal wires less than .02 inch in diameter. Other whistling sounds are not usually associated with feeding activities, and may be a form of communication.

Harbor porpoises' main enemies are large sharks and killer whales. Northwest coastal tribes occasionally harvested these mammals for a highly prized meal, but now man only incidentally takes them in salmon gill nets. Until several decades ago this small porpoise was Puget Sound's most common cetacean. Today they are limited to the San Juan Islands area and the Strait of Juan de Fuca. Some biologists think greater vessel traffic in the Sound may have been just too much for these timid sea mammals.

Dall's Porpoise

Passengers on the numerous ferries that ply the waters of Puget Sound and the San Juan Islands sometimes are entertained by playful Dall's porpoises that gracefully ride the bow waves of the bulky craft, and then quickly vanish into the depths. Puget Sound fishermen and sightseers in smaller boats also enjoy these popular marine mammals, sometimes maneuvering their craft in an effort to attract a porpoise.

The Dall's porpoise (*Phocoenoides dalli*) is one of the swiftest members of the porpoise family,

able to reach speeds of up to 30 miles per hour. A black body with a large white patch on the belly and flanks makes them distinctive, although they sometimes are confused with young killer whales and the smaller harbor porpoises. Their dorsal fins are triangular and often white-tipped. Males average six or seven feet long and 320 pounds, slightly larger than females. Though they usually do not leap completely out of water, the rooster tail they produce when surfacing rapidly to breathe is easily identifiable.

Dall's are the second most common porpoise along the Northwest Coast. Unlike their harbor porpoise relatives that frequent shallower bays, Dall's porpoises generally make their home in deeper waters. Except where there are deep channels close to shore, these mammals rarely are observed from land, but schools may be seen by mariners venturing farther out to sea.

They feed at night on herring, hake, rockfish and jack mackerel, as well as an occasional squid, locating food with their well-developed sonar system. Although these porpoises have plenty of teeth, they are designed for seizing and holding food rather than chewing it, requiring them to swallow their prey whole. Because Dall's porpoises have a relatively small amount of insulating blubber to protect them from their cold, watery environment, they must produce a great amount of body heat. To meet this energy demand their metabolic furnace is stoked daily with an amount of food that exceeds 10 percent of their body weight.

The killer whale is the Dall's only known natural enemy. Along the Northwest Coast these mammals sometimes were caught by natives, but commercial whalers gave them scant attention. In the northeastern Pacific several thousand are taken each year by the Japanese as an incidental part of the high seas salmon gill net fishery.

The Dall's porpoise is more tolerant of human disturbance than the shy harbor porpoise. As a result, it may, to the delight of recreational boaters, be replacing its more timid cousins in parts of Puget Sound where boating and commercial shipping have increased greatly.

Left: Dall's porpoise viewed from above off the bow of a boat. James T. Harvey photo.

Douglas squirrel with sitka spruce cone. Thomas W. Kitchin photo.

Mountain Beaver

If people ever decide to name a "regional mammal" (not animal!) of the Northwest, the mountain beaver would be a prime candidate if it could be seen or photographed. This species is most abundant in the oceanside forests of Washington and Oregon. Possibly the 19th century Nisqually Indians recognized the antiquity of the mountain beaver, as one of their mythological traditions held that it was the first animal created.

The original account of the mountain beaver appeared in the notes of Lewis and Clark in December 1805. At their Columbia River winter quarters in Fort Clatsop, the explorers bought several skins of an unfamiliar animal from the Chinook Indians. When questioned about the origin of the furs, the Indians answered, "she-wal-lal," which the white men mistook as the name of the furbearer. The Indians, however, were referring to the blanket-like robe made of the skins.

Though they don't look much like the real beaver, they do have certain beaver-like habits. They divert streams into their tunnels, gnaw bark from trees and cut off small trees or their limbs, all of which helped earn them their common name.

The reddish-brown mountain beaver, locally known as the boomer, is well-described as a muskrat without its tail. They average 10 to 14 inches long, weigh between one and three pounds, and have a short, stubby tail. A tiny white spot decorates each cheek below the ear. Hearing and eyesight are both poorly developed, whereas the senses of smell, touch and possibly taste guide the creature safely through its nightly activities.

Mountain beavers (*Aplodontia rufa*) usually live underground, using the long claws on their front feet for digging. Streamsides in coniferous forests with nearly impenetrable jungles of thimbleberry, blackberry, and salmonberry are likely candidates for their tunnels. Entrance holes to their complex burrow networks often are conspicuously marked by surrounding dirt piles. Tunnels are close to the surface and include a nest chamber with storage facilities. Some dead-end tunnels also are dug and used as refuse areas for disposal of excrement and garbage.

The nest has a base of coarse vegetation and is lined with fine, dry plant material. Mountain beavers dig storage chambers for food, which is cut while fresh and gathered in piles. Sometimes a "haystack" of wilting skunk cabbage may be found on or under a log near the entrance. The wilted material is later transported into storage chambers where it is either eaten or kept, remaining moist because of the burrow's high humidity.

The average home range of mountain beavers measures a diminutive one-third acre. Eighty percent of individuals never venture farther than 25 yards from their home. Generally solitary, these creatures are also territorial, defending their nest sites against any intruders. Encounters are frequent since small home ranges overlap.

Mountain beavers are vegetarians. Swordfern, bracken fern and salal are staples throughout the year, but almost any plant growing close to the burrow will be used. Most of the feeding is done at night and the early morning hours. This mammal stays active even in the winter, burrowing through the snow when necessary to reach the bark of young conifers.

The breeding season occurs in January and February, and after an approximate 30-day gestation period a litter of three to five helpless, scantily haired and blind young is produced. Little ones grow very slowly and depend on their mother's milk for at least two months, thus limiting her ability to produce litters to once a year.

Burrows of mountain beavers are favored by several other mammals, such as snowshoe hares, long-tailed weasels, minks and spotted skunks. The last three species probably helped to vacate the site since they are known predators of these animals. Another unwelcome guest in mountain beaver nests and burrows is a rare, giant flea. It is reputed to be the largest flea in the world (3/8 of an inch long) and is found exclusively on or near mountain beavers.

In the Northwest the mountain beaver is sometimes considered a nuisance by gardeners when their flower and vegetable beds are ravaged by its burrowing. The destruction is not restricted to gardens, however. Mountain beavers can cause considerable damage to conifer seedlings, which they destroy by gnawing or by clipping the stems. They also kill trees an inch or more in diameter by girdling them around the base.

By the late 1970s concern about mountain beaver damage led to formation of the Northwest Forest-Animal Damage Committee, with a mountain beaver subcommittee. A study conducted by the group found that mountain beavers caused damage on about 275,000 acres of forest land, primarily to Douglas fir stands. Major problem areas were in the Olympic Peninsula south to Willapa Bay and eastward to the Puget Sound Trough in Washington, and in the Coast Range of Oregon.

A most effective means of controlling mountain beaver damage is to install plastic mesh tubes around the stems of seedlings. An alternate method is trapping problem animals, but this is not cheap, costing $10 to $15 per acre. As with all animal-damage control efforts, a little public relations work is advisable. Some people objected to trapping mountain beavers because they thought they were real beavers. Foresters have been encouraging use of the other common name, boomer, to help the public realize that the problem is caused by a small pest, not a valuable furbearer.

Yellow-Bellied Marmot, Hoary Marmot and Olympic Marmot

The three Northwest mountain-dwelling marmots considered here inhabit some of the roughest, coldest and most difficult terrain in the contiguous states. Boulder-covered slopes and adjacent alpine meadows of the high country in the Rocky Mountains of northern and central Idaho and western Montana, as well as the North Cascades of Washington, offer a suitable environment

Hoary marmot dines on arnica flower in the Cascades. Thomas W. Kitchin photo.

for the country's largest member of the squirrel family—the hoary marmot (*Marmota caligata*). A few thousand of its close relatives, the Olympic marmots (*Marmota olympus*), inhabit 700 square miles of similar terrain in Washington's Olympic National Park. More common is the yellow-bellied marmot (*Marmota flaviventris*), which can be spotted in basaltic outcroppings and other rocky and dry areas in the lower valleys and foothills of central and eastern Oregon and Washington, and in the mountains of western Montana and Idaho. It occupies areas from near timberline to 11,000 feet, wherever the dominant hoary marmot is absent. A few groundhogs, or woodchucks (the same variety as Pennsylvania's Punxsutawney Pete of

Groundhog Day fame), have made their way into northern Idaho and northeastern Washington.

All three marmots are large, heavy-set rodents with bushy tails. The hoary marmot is silvery gray above with a whitish underside, making it well camouflaged against the gray granite rocks it seems to like so well. It also has distinctive black and white markings on its head and shoulders, and wears a set of black fur "boots" on its feet. Olympic marmots are uniformly dark brown, though their coats may become bleached to a lighter yellowish-brown by the end of summer. The yellow-bellied species sports yellow underneath, while the white-tipped hairs on its back give it an overall grizzled, yellowish-brown appearance.

Whitish spots decorate its forehead between the eyes and the buffy colored patches on its neck and shoulders are distinctive from the hoary marmot.

In true groundhog fashion, all three species inhabit burrows. Those of the yellow-bellied marmot are almost always under rock piles or in crevices between boulders, although roadside banks and log piles also are used. Hoary marmots prefer to dig under large boulders in building their dens, while the up to five-acre interconnected tunnel systems of the Olympic marmots often are excavated directly below the meadow that provides their daily meals.

All are generally vegetarians, spending large portions of each day feeding on green plants, including tender shoots of young grasses, legumes and other succulents dotting surrounding high country meadows. Occasionally, animal foods such as insects might be consumed. They spend winter months in hibernation and, depending on location and elevation, may stay in this inactive state for up to seven months. To survive such a long dormancy they must accumulate a thick layer of fat, which may boost an animal's weight by as much as 50 percent during the short alpine spring and summer.

Marmots mate soon after emerging from winter quarters, and about four weeks later two to six young are born. Yellow-bellied marmots have one litter each year, whereas both other species have one of the slowest reproductive rates of all rodents, with a female producing as few as six, and normally no more than 18, offspring in her seven-year life.

Cooperation among individuals within their social colonial groups seems to help the survival of all. With many pairs of vigilant eyes, an intruder seldom goes unnoticed. At the first sign of danger at least one colony member produces a shrill whistle, which sends less watchful foraging members bounding for the safety of their burrows. In any closely knit society, differences of opinion are bound to develop. Aggressive interactions among hoary or Olympic marmots may be contained by play fighting. This is especially popular among

Right: Aptly named yellow-bellied marmot. Thomas W. Kitchin photo.

Left: Greeting, marmot style—two hoary marmots touch noses. Thomas W. Kitchin photo.

Below left: Amidst avons and wallflowers, an Olympic marmot surveys its Olympic National Park home from Hurricane Hill. Thomas W. Kitchin photo.

the younger members of the colony, which wrestle each other or, standing on hind feet, try to push each other over backwards. This activity may help to establish a stable order of dominance in the colony by the time the young mature. A highly developed greeting system also helps reduce tension. When two marmots meet, they usually rub their noses together and touch mouths. This is often followed by friendly nibbling on the other's neck and ears.

Although marmot burrows are several feet long, an occasional bear might try to dig up the still hibernating individuals during the spring. Other enemies include eagles, hawks, cougars, bobcats, coyotes and badgers. In the basalt cliffs along the Columbia River near larger cities, excessive recreational shooting of yellow-bellied marmots is turning this normally diurnal animal into a nocturnal one. Since marmots living in remote rock outcrops are of no nuisance to man, using them for target practice is a questionable activity. Marmots are rarely eaten nowadays, and the pelts are virtually worthless in the fur market. Years ago, however, yellow-bellied marmot skins were made into extremely durable robes by the Shoshone Indians of Idaho.

Columbian Ground Squirrel

The life of a Columbian ground squirrel is seldom secure. When foraging above ground they are threatened by such four-footed creatures as coyotes, foxes and bobcats. Many feathered predators also are quick to seize such an opportunity. After daily activities are completed and the animals retire into their extensive burrow systems, weasels and snakes may pursue them there. As if this were not enough, badgers and bears may try to unearth them with their powerful claws.

Columbian ground squirrels (*Spermophilus columbianus*), first described by Lewis and Clark near Kamiah, Idaho, are short-legged mammals with a grizzled buff upper body. They have a distinct reddish-brown color on the face, front legs and underparts. Body length averages 13 to 16 inches and most individuals weigh between 12 and 18 ounces. The tail is relatively long (up to five inches), dark and bushy. Their ears are small and rounded, the eyes large. Generous cheek pouches are used for storing food items for later transfer to the burrow.

These ground squirrels inhabit mountain meadows, foothills and rolling grasslands in western Montana, northern and central Idaho, eastern Washington and northeastern Oregon. They are vegetarians to a large degree, feeding on grasses, leaves and stems early in summer. Later, various fruits and seeds are consumed, consistent with their genus name, which means "seed loving." Most of the animal matter they eat is composed of such insects as grasshoppers, beetles and cicadas. In places where sagebrush grassland has been converted to alfalfa, oats or rye, the rodents may move in and become pests. In eastern Washingonton they are particularly bothersome to farmers raising winter wheat. As much as one-third of the crop may be destroyed by these uninvited dinner guests, although losses generally are not so significant.

The squirrels fatten up during summer and may enter summer hibernation by late July to avoid heat and lack of water. They may reappear for a few weeks in late summer only to go into hibernation by early September. The animals emerge in April, quickly mate and may be tending four or five youngsters within a month.

Red Squirrel and Douglas Squirrel

These two species of tree squirrels often are referred to as pine squirrels because of their preference for living in evergreen forests. Though they are similar in size, both measuring about 10 to 15 inches long (including a four- to six-inch bushy tail) and weighing five to 10 ounces, they have different coloration. Red squirrels (*Tamiasciurus hudsonicus*) have grayish-red backs with white undersides, while the Douglas squirrel (*Tamiasciurus douglasii*) is dark grayish-brown with orange or yellow underparts. In both species a black line along the side, especially noticeable in summer, separates the colors. The Douglas squirrel, or chickaree, inhabits the western half of Washington and the western two-thirds of Oregon. Their

Left: Columbian ground squirrel. Bruce Pitcher photo
Above: Western Washington "peanut butter bandit,"
Douglas squirrel (chickaree). Thomas W. Kitchin photo.

The red squirrel, popular campground entertainer. Bruce Pitcher photo.

Beaver

Many a trout in the Northwest enjoys a tiny lake behind a dam built by one of the few mammals besides man that profoundly alters the natural environment for its own benefit. Also within the pond there is likely to be a large, dome-shaped pile of sticks somewhere near the shoreline; the accommodations of the skillful engineer. A flyfisherman wading to such a site on a cool summer's eve may be startled to hear a loud "whack" on the water's surface, a message delivered by the large, canoe-paddle shaped tail of a beaver to warn all other creatures, including trout, of an intruder.

Beavers (*Castor canadensis*) have not always had to worry about human invasions. Almost 40 million years ago, *Stenofiber,* the first known forefather of the beaver, roamed the wetlands of this continent. Much later during the Pleistocene Epoch, the beaver shared his favorite haunts in North America with a cousin, *Castoroides ohioensis,* until the end of the last Ice Age some 12,000 years ago. This giant measured 10 feet in length and weighed an incredible 600 pounds, resembling the modern black bear in size. The modern beaver is a mere four feet long from the end of its nose to the tail tip. It rarely exceeds 60 pounds, but holds the title as the New World's largest rodent.

Well-equipped for their aquatic way of life, beavers appear uneasy and clumsy on land. Their silver-gray underfur is designed to trap air for insulation when swimming, while the long reddish-brown guard hairs aid in keeping water from penetrating even that far. A beaver's hind feet are fully webbed and can propel the creature through water at speeds up to six miles per hour. When the animal dives, remaining submerged for as long as 15 minutes, valves close its ears and nostrils, while the eyes are protected by clear membranes. The mouth is sealed by flaps of skin, but its large, dark orange incisors are left exposed so the beaver may carry twigs under water.

Beavers can be found in all suitable Northwestern aquatic habitats from sea level to 12,000 feet in

Right: Beaver has its fill of tender willow branch. Alan Carey photo.

red cousins, on the other hand, range over most of Idaho, western Montana, eastern Washington and northeastern Oregon. In Washington's North Cascades the two species occasionally interbreed.

Choice food items are various conifer seeds, especially spruce. Each individual collects up to 14,000 cones during the summer and caches them for future use, often in a moist site to keep cones from opening. Seeds from such cones germinate more dependably than those picked directly from trees, and squirrel caches often are important seed sources for tree farms in the Northwest. To the dismay of many livestock owners, the industrious squirrels may completely fill a large horse or cattle watering trough, leaving little room for water.

Both relish mushrooms, which they lay on tree branches or logs to dry before tucking them away in some sheltered place. Fortunately the little animals can distinguish a "Destroying Angel" Amanita mushroom from a choice morel. The

feed cache usually is close to a stump, hummock or other suitable feeding site on which the squirrel sits while enjoying its meals. Discarded cone scales form a midden, which may be up to a yard tall and five yards in circumference, with the bottom refuse decades old. When seedbearing cones are scarce, pine squirrels rely on winter buds of evergreens, sap from maples and birches and even bark.

Although these tree squirrels seldom live more than five years, mortality due to predation is light. Several animals, including red-tailed hawks and goshawks, red foxes, martens and fishers occasionally include them in their diets, but no predator relies on them for sole sustenance. Red squirrels are notorious for moving into vacant summer homes where they can do significant damage. More often both the red squirrel and chickaree are appreciated for the impromptu, live entertainment they provide Northwest campers.

Handiwork of an industrious beaver. Alan Carey photo.

elevation. Their profound impact on the environment makes their presence easy to detect. They build their dams on rivers and streams to ensure an adequate water depth for their homes. A "busy beaver" can perform prodigious building feats. One undulating dam in Montana measured 2,140 feet long! More typically they are 10 to 50 feet in length, and 3 to 5 feet high. Dams usually are built with 1- to 4-foot sections of aspen, willow, birch or cottonwood limbs, but a beaver uses whatever is available. In some dry areas devoid of trees, sagebrush is the standard building material. In the Oregon Coast Range small one-foot-high beaver dams built by the animals piling up stream gravel are common. Dams are made more watertight by the addition of mud, which the beaver carries from the pond bottom between its forepaws and chin. Although it looks like an ideal tool, a beaver does not use its tail to plaster mud on the dam, contrary to the cartoon stereotype.

Once a stable pond three or more feet deep has been formed, the beaver turns its attention to completing a sturdy lodge, also made of tree limbs and

Northwesterners who want a firsthand feeling for the history of mammals do not have to travel very far. The John Day Fossil Beds National Monument in northcentral Oregon contains one of the longest continuous records of mammalian evolution in the world. The fossil beds and nearby John Day River are named after a member of the Pacific Fur Company's Overland Expedition to the mouth of the Columbia. In 1812 Day and his dwindling number of colleagues were too busy surviving in the wilderness and battling Indians to have worried about fossils, assuming they even knew such things existed in the region. In fairness, when the 14,000-acre National Monument was established in 1974, it should have been named in honor of pioneer missionary and evolutionist, Thomas Condon.

Condon's calling found him arriving to set up his practice at the Dalles in 1862. Soldiers at the local army post heard of his interest in natural history and alerted him to the numerous fossils in the hills along the John Day River, a hundred miles to the southeast. Condon began making collecting trips and, recognizing the immense fossil resource, informed leading geologists of his discoveries. Before long, titans of American vertebrate paleontology were bickering over who would have the honor of describing each new mammal species discovered.

The work of these and later paleontologists revealed that the fossil beds' five major rock formations cover a time span of from more than 50 million to less than five million years ago—most of the Age of Mammals.

Fossil plant leaves and seeds showed that a subtropical Central Oregon grew palms, figs and avocados 40 million years ago. Ancient rhinos and tapirs roamed the swamps, along with huge blunt-horned mammals called titanotheres.

John Day Fossil Beds

Sheep Rock in the John Day Fossil Beds National Monument is composed of weathered layers of volcanic ash that date back to the formation of the Cascades. Fossils from these layers of time record millions of years of mammalian evolution. According to Ted Fremd, curator of the monument's museum, the deposits constitute a "world class site for the Age of Mammals." John A. Alwin photo.

Thirty to 20 million years ago the climate changed to warm-temperate. Remnants of giant pigs, camels, saber-toothed cats, rhinos, oreodonts, peccaries, wolf-like carnivores and primitive opossums were found in John Day formation rocks. Ten million years later, primitive raccoons, weasels, mountain beavers and pronghorns appeared on the scene. By five million years ago, the rising Cascades produced a distinct rain shadow effect in John Day country. By this time relatives of the modern pronghorn, large camels, mastodons, a primitive coyote and bears frequented the much warmer, drier environs.

At the Painted Hills Unit, eroded volcanic ash has acquired attractive hues of red, pink, buff and gold. Fossil remains of Stenofiber, *the first known forefather of the beaver, have been unearthed from these deposits. John A. Alwin photo.*

mud. After piling a heap up to eight feet high, one to three underwater entrances are built into the stack. Tunnels converge at living quarters hollowed out within the center of the structure. To allow fresh air to enter the nest chamber, the top of the lodge is not coated with mud. Beavers living along swift rivers usually dig a bank burrow instead of trying to tame a large flow. As if all this engineering were not enough, beavers dig canals up to 700 feet long to connect their foraging areas with the home pond. These watery highways are used to float food items and building materials to the main pond.

Considering their engineering feats and seeming foresight in storing winter food, it is easy to understand why numerous Northwest Indian legends center around this extraordinary mammal. According to Ross Cox, North West Company clerk and fur trade chronicler, the Flatheads thought beavers were a fallen race of Indians. Traditions held these Indians-turned-beavers had angered the Good Spirit, who transformed them into the large rodent's form. According to legend, male beavers retained human speech and had been observed conversing in council.

The beaver is active year-round, feeding on succulent plants during summer, and aspen bark, the staple in the diet. In autumn, green twigs and branches of aspen, willow and other deciduous trees are laid in the bottom of the pond close to the lodge as a supply for winter. If the pond freezes, the beaver has only to swim a short distance from its home to the larder and pick up a ready-made meal. Inside the lodge, the bark is gnawed off the branch in the same way an Iowa farmhand polishes off an ear of corn.

Below: Contrary to the cartoon image, the beaver does not use its tail to plaster mud on a dam. Alan Carey photo.

Beaver, Audubon lithograph. Audubon, Quadrupeds. *Photo from the Special Collections Division, University of Washington Libraries.*

Young beavers enter the world with open eyes and a complete fur coat. They develop into excellent swimmers by the end of their first week, but remain with their parents for two years. During this time they help with constant repair work on the dam, and other chores around the pond. After attaining maturity in the second spring of their lives, they seek a suitable site for a new colony. During these travels they may encounter scent mounds, mud piles on which an oily substance called castoreum is deposited by all resident beavers in a colony. Castoreum is produced by two castor glands located in the beaver's groin, and is used for marking the boundaries of already occupied territories.

Exploration and early settlement of the inland Northwest largely was due to the white man's pursuit of the beaver. In today's world of polyester and nylon it is hard to fathom how a large rodent could have been so important in the lives of men and women, until we remember that human vanity can be a powerful driving force. Beaver pelts may be made into luxurious clothing and, more importantly, the downy underfur can be chopped to make felt for once fashionable hats.

Beavers were abundant throughout North America when the white man arrived. During early American Colonial times, large numbers of pelts were shipped from the Colonies to Oriental markets. This outlet declined dramatically by 1675, but the European market did not immediately pick up the slack. At that time European beavers still could be obtained, and their fur was considered superior to American beavers for hatmaking.

Over the next 100 years, Europe's beavers virtually were exterminated, and overseas sales of American pelts grew sporadically. In the late 1790s, demand increased, and fur shipments from America almost doubled by 1808. One author noted this should have been called the "fiber trade" instead of fur trade, because beaver for hatter's felt was by far the most important commodity. As the price of pelts increased, both Great Britain and the fledgling United States turned covetous eyes toward the Northwest's rich fur resources, which had been hinted at by sea captains who had traded with Native Americans on the West Coast.

The inland Northwest was still unchartered territory when President Thomas Jefferson received funding from Congress in 1803 to authorize a transcontinental exploration trip. More was involved than satisfying man's seemingly natural quest to seek what lies beyond the sunset. Among several major objectives were ascertaining the region's potential for fur production and laying the groundwork for challenging the British for control of that commerce. Lewis and Clark's epic journey of 1804-06, the first American exploration venture into the American Northwest, took them to the mouth of the Columbia River and back. The Corps of Discovery returned with reports that at least part of the Northwest was "richer in beaver . . . than any other country on earth." The rush was on to this fur bonanza.

The Canadian North West Company (NWC) took an early lead in the race to tap the region's beaver potential. Operating out of today's Alberta, Canada, Nor'Wester David Thompson led that company's advance south of the 49th parallel and into the Northwest. By 1810 the NWC had four trading posts, or "houses," in northwest Montana, northern Idaho and northeast Washington. Other posts followed but, so too, did competitors.

The venerable Hudson's Bay Company (HBC), the NWC's chief rival in British territory to the north, tested the Northwest waters with an 1810-11 trading expedition to western Montana's Flathead Lake area, but opted to stay out of the Northwest trade at that time. American competitors proved only marginally more persistent. In 1811 John Jacob Astor's Pacific Fur Company entered the scene with establishment of Fort Astoria at the mouth of the Columbia. Subsidiary inland posts followed, but the War of 1812 and likelihood of British attack forced the Astorians to sell out in 1813. Overnight, Fort Astoria became the Nor'Westers' Fort George.

The NWC enjoyed a near-monopoly in the Northwest fur trade until their 1821 absorption by the HBC. The senior HBC (some say the initials really stood for "Here Before Christ") wasted little time

Recapitulation of — Furs Shipped 1830

			New Caledonia	Fort Vancouver	Fort Langley	Coasting Trade	Fort Colvile	Fort N. Forces	Thompson's River	South'n Express	Snake Expedition	Total	New Caledonia	Fort Vancouver	Fort Langley	Coasting Trade	Fort Colvile	Fort N. Forces	Thompson's River	Total
					Outfit 1829										**Outfit 1830**					
Badgers														5						5
Bears	Black	Large	58		2		26	3	11			100	3	15	1	1	2		4	126
		Small	16				4	2				22	5	9		2	1		8	47
	Brown	Large	1				18	6				25		7		1	2		3	38
		Small	2				5	3				10	2	8			5		3	28
	Grizzle	Large	1				17		6			24					1		1	26
		Small					1					1							1	2
Beaver	Large		4221	1459	889	47	2078	215	497	2018	1091	12715	172	1484	147	156			44	14718
	Small		1908	766	316	3	1134	159	205	222	204	4917	35	350	77	54	13		18	5464
	Large										468	468								468
	Small	Damag'd									53	53								53
	Coating	lb	561	57	4		20	23	10	16	20	711	72	125	8	4	14		30	964
Fishers			98	3			244	18	53		1	419	34		2		42		10	507
Foxes	Cross		49	8			15		32			105	71		6		48		15	245
	Red		43	11			48	9	55			166	24						4	194
	Silver		37	3			1		12			53	47						5	105
Lynx			801				19		67			887	243	2		22			76	1235
Martens			2204			1	331	7	565			3107	570	19	52	41	19		146	3934
Minks			98	8			249	41	94			510	3	40	38	11	128		22	752
Musquash			1768	501	25	3	5342	140	2074		4	9857	166	271	120	43	120		535	11117
Otters	Sea	Large		4								4			2					5
		Small		4								4		7	2					13
	Land	Large	207	530	378	22	192	49	90	236	90	1794	18	542	64	39	17		15	2470
		Small								24		24				15			5	44
		Damag'd						2		66		68								68
Raccoons							3					3								3
Swans														1						1
Wolves			10				7					17	3	5						25
Wolverines			26				10		3			39	21	1			2			43
Castorum		lb	252				54		25			331		58					21	418

"Recapitulation of Furs Shipped 1830" from the Fort Vancouver Account Book. Total count of furs from each district and other trading ventures (including the Snake Expedition) of the company's Columbia Department for the two trading years, or "outfits," of 1829 and 1830. Courtesy of the Hudson's Bay Company Archives, P.A.M., B.223/d/27, fos. 6d-7.

A senior Peter Skene Ogden, one of the Northwest's legendary fur trade personalities. Oregon Historical Society photo, neg. no. 707.

in integrating this fur-rich region into its continental operation. Fort George was abandoned in 1825 in favor of newly constructed Fort Vancouver, farther upriver. From this British bastion on the Columbia River, the HBC orchestrated one of its most productive fur districts.

The Company's potentially most serious opposition came from the east in the form of American trapping parties. To tap that region and eliminate beaver, thereby making it unattractive to the Americans, the HBC relied on the famous Snake Country Expeditions. These horse-mounted, winter forays to trapping grounds in eastern Oregon, southern Idaho, southwest Montana and adjacent areas departed one of several Northwest posts each fall under now historic fur trader personalities—Alexander Ross, Peter Skene Ogden and John Work.

Their objective was clear, to devastate beaver populations in those peripheral areas in what one fur trade historian has called a "scorched stream" policy. Even hard-bitten Ogden was not oblivious to the long-term impact of such ruthless trapping. In April 1829 while camped near the South Fork of the Owyhee River, he recorded in his journal, "It is scarcely credible what a destruction of beaver by trapping at this season, within the last few days upwards of fifty females have been taken and on

The Hudson's Bay Company's Fort Vancouver as it appeared to an artist with the federal government's 1853-1855 Northern Railroad Survey. Even in these waning years of the fur trade this post remained an important trade and supply center for regional settlers, including Willamette Valley residents, and even supplied California gold miners. At the time the fort was described as "enclosed by a stockade of two hundred by one hundred seventy-five yards, twelve feet in height, and is defended by bastions on the northwest and southwest angles mounted with cannon...At some distance there is also a village of fifty or sixty cabins occupied by servants [company employees], Kanakas, and Indians, and a salmon-house on the bank of the river." At the time Peter Skene Ogden shared command of this one-time capitol of the Northwest fur trade. Color lithograph from Reports of Explorations and Surveys...1853-5.

an average each with four young ready to litter. Did we not hold this country by so slight a tenure it would be most to our interest to trap only in the fall, and by this mode it would take many years to ruin it."

Although other furs were sought, beavers were central to the HBC's trade. In fact, other furs and all trade goods were valued relative to one large prime beaver pelt, which was assigned a worth of one Made Beaver (MB). For instance, at Spokane House during the 1824-25 season, a muskrat equaled 1/10 MB and a marten 1/3 MB. Among a wide array of available trade goods, Indians had to "pay" four MB for a pistol, two MB for each pound of twist tobacco, one-half MB for small scissors and four MB per pound of gun powder.

Just as vanity and politics had served to decimate beaver numbers, they also provided some relief. European fashions changed in the 1830s when silk, wool and other furs began replacing beavers as hatter's raw materials. Evidently the South American coypu (nutria) was one substitute. In 1841, the general London fur market collapsed, and when it recovered, beaver was no longer desirable or available. Dwindling supplies of beaver left little for American and British trappers to fight over, and the British-American agreement in 1846, setting the boundary between the United States and Canada west of the continental divide at the 49th parallel, limited the future of British trapping in Northwest America.

The advent of new game protection laws in most states during the late 19th and early 20th centuries allowed the beaver to stage a comeback. In much of Oregon, Montana and Idaho, the prolific mammals recovered naturally, while reintroductions helped beavers gain a foothold in still other sections of their former range.

Even with liberal trapping seasons and annual harvests of up to 60,000, Northwest beavers can become pests when their dam and pond building activities conflict with human desires for commercial forests or dry backyards. According to Frank Newton, statewide furbearer biologist with the Oregon Department of Fish and Wildlife, 514 reports of beaver damage were received by the department in the 1984-85 fiscal year. Even though the Beaver State may have fewer than 70,000 beavers, compared to an estimated one million in the early 19th century, "complaints are on the increase," Newton reports. If removing beavers is a primary management goal, one thing is certain—dynamiting a dam does not work. A beaver loves a challenge, and always will come back.

The "O.S.U. Beaver." Used with permission of the Office of the President, Oregon State University.

Muskrat and Nutria

If you inherit a fur coat with a label marked "Hudson Bay Seal," don't be ashamed to wear it out of fear of encouraging marine mammal overexploitation. The fur industry decided that this somewhat misleading trade name has a better ring to it than does the name of the fur's original owner, a muskrat.

Another source of misunderstanding about the muskrat is trappers and fur buyers who call it a rat. It does look a little like a giant aquatic rat, but it is not closely related to such an offensive creature. A more accurate name taxonomically would have been muskvole. The first part of the muskrat's name is to the point, however, since these animals do have two musk glands on either side of their lower abdomen. The glands attain their highest development during the breeding season, and are used to inform each sex of the other's whereabouts. Man learned long ago to apply the oily glandular secretion on his traps to lure unsuspecting, amorous individuals.

Muskrats (*Ondatra zibethicus*) are semi-aquatic, mostly nocturnal rodents that seldom venture far from water. Marshes, swamps, rivers, streams and ponds as well as man-made structures, like irrigation canals and reservoirs are suitable habitat, as long as they provide some aquatic vegetation. The 'rat is well adapted to this environment, with its small eyes and ears, large hind feet with partially webbed toes and a long, scaly tail that serves as a rudder. Its compact body is covered by dense, fine underfur and long, coarse guard hair, which gives the animal a glossy appearance. These qualities of the pelage make it an excellent insulator against cold weather, and a valuable fur for people.

Muskrats feed on aquatic vegetation, especially cattails, and occasionally resort to terrestrial plants, including skunk cabbage and ferns. If a cultivated field lies adjacent to their watery homesite they gladly partake of alfalfa, corn and other crops. Small animals, including snails, tadpoles and crawfish are taken. Individuals inhabiting the shore of Puget Sound even feast on marine mussels.

Aquatic vegetation also is used to build the conical houses so conspicuous in muskrat-infested marshes. Plant leaves, stems and roots, as well as mud are used to create a shelter up to eight feet in diameter and five feet high. It rests on the bottom of the pond and has one or more underwater entrances. In areas with running water, these rodents prefer to live in dens dug into the stream bank, with entrances below water level.

Breeding takes place from May through October and, after a 29-day gestation period, an average of six young are born. They are helpless at birth but surprisingly tough, being able to go without food for several days. These traits probably have evolved to assure the survival of at least a few in the hands of a less-than-perfect mother. She tends to lose or trample her infants, and those that survive early childhood are forced to leave before the birth of a second litter. If fate is kind the youngsters learn to swim and dive, and begin to eat green vegetation by the end of the second week of their lives. They grow rapidly and the precocious

Above: In western Oregon and southwestern Washington people sometimes confuse these similar-looking nutria with beaver and muskrat. Tom and Pat Leeson photo Right: Muskrat feeds on tubers. Thomas W. Kitchin photo.

youngsters usually start looking for their own quarters at the age of four weeks.

Muskrats benefit man in many ways. In wildlife refuges, such as Malheur in southeastern Oregon, 'rat populations, through proper management, keep vegetation at the level best suited for waterfowl nesting. The number of trapping permits is carefully controlled so that 'rats do not "eat out" all the vegetation or, conversely, let plant growth become too rank.

These rodents also are one of the most important furbearers in the Northwest with an annual harvest of about 250,000 individuals. Idaho consistently accounts for about half the harvest. Record prices in 1981 of about $5.60 per pelt meant that year's state take of 149,000 'rats had a value of over $750,000! Interestingly, within the Northwest it is the semi-arid and treeless plains sections with extensive irrigated farmland that produce the largest number of muskrats. The dry, irrigated Snake River Plain of southern Idaho leads that state, while eastern Washington's parched Grant County, home of the sprawling Columbia Basin Irrigation Project, is the Evergreen State's 'rat factory.

In western Oregon and southwestern Washington an animal similar in appearance to the muskrat, though considerably larger (5 to 25 pounds), might be seen. This is the nutria (*Myocastor coypus*), imported by an estimated 1,000 fur farmers in the 1930's. When the industry collapsed after World War II, many of these farmers simply turned their stock loose, despite laws prohibiting release.

The introduced species became an immediate nuisance to farmers, vandalizing their crops and canals. Neither are they loved by waterfowl refuge managers in the Willamette Valley, where they damage earthen dikes. Lately, trappers in Oregon and extreme southwestern Washington have been getting their share of the several million dollars paid each year for nutria pelts, thus turning a pest into a coveted furbearer. The Northwest's annual take of 16,000 pelts, almost all from Oregon, generally fetch a better price than muskrats. Nutria fur prices are well known for their wide fluctuations, however, and in another 10 years the South American rodent may once again be relegated to the "complete pest" list.

Porcupine

Frantic barking followed by agonized howls and whimpering produces dread among dog owners. A minute later "Duke" returns, tail tucked between his legs, and gives his master a puzzled look with a face newly decorated with whiskers. This is probably the most common, and aggravating, way humans encounter the second-largest rodent on this continent. The porcupine's scientific name, *Erethizon dorsatum*, means "irritable back," an apt appellation for an animal that certainly produces its share of irritations.

Though this ambulatory cactus has some 30,000 one- to four-inch-long quills adorning its rump, back and sides, it is generally cautious and unaggressive, preferring to retreat into a rocky crevice or up a tree rather than confront its enemy. When cornered, however, it lowers its head between its front legs, turns its rear toward the

opponent and, with a few well-aimed slashes of its strong tail, drives home a couple of dozen barbed quills. Quills are modified hairs, barely attached to voluntary muscles underlying the skin, and are easily detached at the slightest contact. Porcupines do not, as folklore suggests, throw their quills at an antagonist.

Half the size of a beaver, the heavy-set, yellowish-black porcupine can weigh as much as 40 pounds. It has small eyes and weak eyesight, but the deficiency caused by poor vision is offset by keen hearing and a good sense of smell, which are more useful in its nocturnal lifestyle. Although the porcupine is quite at home on the ground, it is an excellent climber and has features modified for an arboreal life. The soles of its feet are covered with small, fleshy knobs. These long, heavy, sharply curved claws both help to grip tree bark. The relatively short, heavy tail also is used in the climbing effort, acting as a prop for the animal to lean on while making its slow ascent.

Porcupines occupy most suitable wooded habitats throughout the Northwest, generally preferring the more open coniferous forests, though they sometimes can be found in sagebrush flats miles from the nearest tree. Before the white man's arrival, these rodents were seldom seen in western parts of Washington and Oregon, although Indian tribes east of the Cascades commonly decorated their garments with porcupine quills. The dense virgin forests in the west described by one early explorer as "thick as the fur on a dog's back," evidently did not suit the "quill-pig's" way of living. Logging opened up the habitat sufficiently, and now the porcupine is found west to the coast and apparently still is increasing in numbers.

Solitary animals by nature, they usually spend their days resting in hollow trees and logs or in underground burrows. Some prefer to loaf in the fresh air of a treetop, especially during warm weather. In general these mammals do not bother to build a true den, and rarely do they carry in bedding materials to soften their sleeping quarters. They are not territorial, having little to lose if

somebody else decides to take over their accommodations.

At nightfall porcupines usually emerge from their hideouts and start foraging to satisfy their tremendous appetites, eating as much as three pounds at one sitting. In summer they feed mainly on green plants, such as skunk cabbage, lupine, clover and various grasses and sedges. Leaves, twigs and catkins of a variety of shrubs and hardwoods also are to their liking. Porcupines do not hibernate, and with winter's arrival, their diets shift to cambium, the inner bark of conifers, preferably pines. If coniferous trees are absent, porcupines gnaw on aspen and fruit trees.

Most breeding activity takes place in October and November when amorous individuals tend to be very vocal, uttering grunts, growls and squeals. The female carries the fetus for about seven months and gives birth to just one offspring in May or June. When born, soft (fortunate for Mom), short quills already are present and they harden within hours when exposed to air. An average lifespan is seven to eight years, though sometimes a Methuselah as old as 11 years can be found.

In spite of the formidable quills, many carnivores will risk a porcupine dinner. Fishers are the most effective predator, and cougars, bobcats and coyotes also relish their flesh. Humans, however, with their traps, guns, poisons and automobiles take a heavier toll on this creature than any of its predators.

Porcupines especially are attracted to anything salty, like doors and floors of cabins, wooden tool handles and saddles that have absorbed perspiration. Their fondness for newly emerged alfalfa shoots and the bark of domestic fruit trees makes them generally unwelcome visitors among farmers. Damage inflicted on commercially valuable timber by girdling the trunk and limbs of trees further decreases their popularity. As part of a porcupine control program in 1961, 24 fishers were imported to Oregon and released to hunt down the rodents. Bounties have been paid for their hides, and there were even signs along central Oregon highways pleading, "Please kill all porcupines." The porcupines may have had the last laugh, however, since sharp quills have been known to make tires go flat.

Above: A porcupine's quills harden within hours of birth. Alan Carey photo
Above right: Porcupines relish anything salty, including a well-used axe handle.
Alan Carey photo   Right: Porcupines have some 30,000 one- to four-inch-long
barbed quills. Bruce Pitcher photo.

Pikas, Hares and Rabbits

Pika. Bruce Pitcher photo.

Pika

Most hikers who enjoy the breathtaking scenery and dazzling array of wildflowers in the Northwest's high country probably have heard a shrill "eenk" or two when nearing a rock slide adjacent to an alpine meadow. This was the special call sounded by a watchful pika to warn his kin of approaching danger. Pikas have extremely keen hearing and excellent eyesight, making it almost impossible to catch them by surprise during their daily activities.

Although pikas are related to rabbits and hares, outwardly they appear very different. Pikas are small, averaging only seven inches in length and weighing but four ounces. These brown, cuddly-looking animals have small and rounded ears with white edges, and no tail.

Pikas (*Ochotona princeps*) live in rugged and remote environments throughout the Cascades and northern Rockies. Most commonly they are found at or above tree line (8,000 to 13,500 feet). Here talus slopes, broken rocks at the base of cliffs, and steep, boulder-strewn hills offer the little creatures both a home and protection from enemies. Pikas are social animals, unlike the usually solitary rabbits and hares. They build complicated pathway systems beneath the rocks, sharing them with all members of the colony. However, each individual has its own, private home after it leaves the family nest.

Alpine meadows surrounding their rocky homes provide abundant supplies of food during summer months. Pikas are herbivores, eating just about anything green, but thistles and lupines are preferred.

Rock rabbits, as they commonly are called, were performing ranching chores long before the first hay was put up for cattle in the Northwest. In late summer and fall the daylight hours in a pika colony are spent in frantic activity. Each individual must stockpile enough food to sustain it through the approaching winter. They cut plants and carry cuttings to a boulder near their home and lay them out to be cured by the sun. Every individual usually has several haystacks, and plants are added daily until each pile measures about two feet in height and contains a bushel or more of hay!

Pika tends a hay pile, its personal supply of winter food. Thomas W. Kitchin photo.

In western Montana the most common plant in hay piles is raspberry; others picked in fair numbers are Oregon grape, mountain cranberry and chokecherry twigs. Pikas in the Washington Cascades select quite a different menu, preferring lupines, vetch and dwarf huckleberries. After plants have cured, the hay is stored under rocks close to the animal's home. Pikas do not hibernate during the long winters in their elevated world, and since they accumulate no extra body fat during the fall, they are entirely dependent on their fur coats for insulation and the stored food supplies for daily meals.

During the May and June breeding season, the animals are fairly tolerant of intruders, allowing other pikas to enter their individual territories. After about 30 days, two to six blind and naked offspring are born in each nest. Within a week the little ones practice walking and begin to call. Their growth is extremely rapid and full size is attained in just 40 to 50 days. Even before that age the young leave their mother's home to claim home territories and begin gathering their own hay piles.

Territorial behavior, which is almost non-existent during the height of the breeding season, becomes more pronounced as the summer passes, with each pika defending both its home area and adjacent feeding grounds from other pikas. The most aggressive behavior coincides with the start of haymaking, which is not a community effort. Instead each individual is out trying to assure his own survival through the winter. Thus, it is easy to understand why any advance toward one's tediously-built haystack by a neighbor triggers an aggressive chase.

Pikas have many enemies, the worst probably the weasel. Other predators include wolverines, martens, coyotes, hawks and eagles. Pikas generally are safe from man's activities, although the occasional backpacker's dog might provide a little extra excitement for some colonies.

White-Tailed Jackrabbit and Black-Tailed Jackrabbit

Jackrabbits and sagebrush generate about the same enthusiasm among most people as do liver and onions. A small quantity now and then adds interest to life, but there definitely is such a thing as too much. Backroad travelers in flatlands east of the Cascades would be hard pressed *not* to collide with one, or several, of the seemingly kamikaze jackrabbits during an evening's drive.

The buff-gray, white-tailed jackrabbit (*Lepus townsendii*) often can be seen amidst sagebrush and bunch grass on the higher plateaus and hills of the semi-arid plains of central and eastern Washington and Oregon, as well as most of Idaho and western Montana. This "rabbit," despite its misleading common name, is the largest hare in North America, weighing up to nine pounds. True to its name, it usually has a pure-white tail, although in some cases the appendage is adorned with a dusky stripe. Its black-tailed cousin, which resembles it in overall coloration, has a black tail stripe extending over the rump. This hare is slightly smaller but is equipped with bigger ears. The range of the black-tailed jackrabbit (*Lepus californicus*) largely overlaps the southern half of the white-tail's home country. It spends most of its time in barren, desert-like areas and sagebrush flats, only occasionally venturing to the lower foothills of the grasslands.

In winter the two hares are easier to distinguish, because only the white-tailed animal changes color, turning white or pale buff. At this season white-tails move down to the more arid sagebrush lowlands, sharing habitat with their black-tailed kin. Here the more abundant black-tailed jackrabbit seems to be outcompeting its rival. This may be due to white-tailed jacks being more conspicuous to predators in these scanty snowfall areas.

Both species are strict vegetarians, feeding on many kinds of grasses and other succulent plants in summer, thus acquiring calories and otherwise

Above left: Early 20th century jackrabbit drive near Aberdeen, Idaho. Wesley Andrews, Oregon Historical Society photo, neg. no. 31972 Left: 'Rabbit drive near Lakeview, Oregon, 1912. D. C. Schminck, Schminck Museum, Oregon Historical Society photo, neg. no. 38153.

hard-to-come-by water. The black-tailed jackrabbit also eats prickly pear cactus and likes to include alfalfa, grains and other cultivated plants in its diet. This habit makes it a nuisance to farmers, especially when man eliminates its natural controls, like the bobcat and coyote. During winter both species depend mainly on woody and dried vegetation, eating buds, twigs and bark of available plants, such as sagebrush and aptly named rabbitbrush.

If jackrabbits are startled, as by an approaching predator, they can leap in bounds 20 feet long and zig-zag wildly at speeds of 35 miles per hour. In short sprints they can reach an amazing 45 miles per hour. When hopping along at moderate speeds, every fourth or fifth leap is exceptionally high, giving the hare a chance to view its surroundings.

Although periodic population explosions suggest jackrabbits were intended to inherit the earth, their numbers are actually on a long slide downward. Throughout their Northwest range, conver-

White-tailed jackrabbit. Bruce Pitcher photo.

sion of vast areas of native sagebrush-grassland steppe into cropland has meant the demise of many a hare. An idea of the former abundance of black-tailed jackrabbits in Idaho may be obtained from the following 1860 quote from a biologist, "These hares are exceedingly abundant on the left bank of Boise' river, where they were so plentiful that a party of sixty men, to which I was attached, subsisted chiefly upon them for a week." One wonders whether jackrabbit numbers rivaled those of the bison, but no one cared to count.

Snowshoe Hare

Cross-country skiers who enjoy nighttime sojourns under a full moon sometimes are startled by a white ghost darting across the shadowed trail, only to disappear as silently as it arrived. The apparition might be attributed to consuming too much hot spiced wine by some, but most likely it was a brief visit by an elusive snowshoe hare, a common animal throughout most of the Northwest.

Snowshoes (*Lepus americanus*) vary from 15 to 20 inches in length. The animal's size seems to depend on the latitude; the smaller individuals usually are found in the southern part of its range, while the larger ones inhabit the northern portions. The typical weight also varies from about two to three pounds. Their 2-1/2-inch ears are smaller than those of other hares, but are larger than the ears of the "true" rabbits. This hare's common name, snowshoe, refers to its large hind feet which are covered with stiff, dense hair. When the toes are spread they create a "snowshoe," enabling them to move easily over the fluffiest surface. The huge footprints left give the impression that a much larger animal is stalking the woods.

Its other common name, varying hare, comes from periodic changes in coloration caused by changing daylengths. In the fall when days gradually become shorter, the hare starts to grow a white coat. In the beginning it is patchy, matching the ground, which is only partially covered with snow. Eventually the entire animal turns "snow" white. Lengthening daylight during the spring

months causes the hare to shed its winter coat and gradually replace it with a brown one to better match the summer colors of its habitat. In coastal lowlands of western Oregon and Washington where snowfall is minimal, the hare remains brown throughout the year.

An adult hare normally flees from danger by running in a large circle with bounds up to 12 feet and speeds approaching 30 miles per hour. Another escape mechanism is called "freezing." After making a short dash at full speed, the animal freezes and seems to totally vanish, an evasive technique young hares master at an early age.

Most snowshoes have two litters per year, each with one to seven young. Like hare size, the number of laverets appears to depend on latitude. Smaller litters are born in the south (an average of three in Oregon), and larger ones occur in the north (an average of five in Washington).

The female does not construct a true nest. Instead, young are born in a shallow depression called a form, situated under cover but on the surface of the ground. Forms usually are in relatively open areas since the adult hares depend primarily on speed for protection. After an amazingly short 36-day gestation period, laverets are born fully covered by hair, with their eyes open, and capable of running within a few minutes of birth.

Wooded areas at nearly any elevation with coniferous forests and brushy undergrowth, and alder thickets along streambeds are preferred habitat. During summers these environments provide the grasses and other herbaceous plants, such as dandelion leaves and clover, central to their diet. Hares also will eat berries and young sprigs of coniferous trees. When their populations are high they can cause considerable damage to seedlings of these trees, as well as fruit trees and ornamentals, which decreases their popularity among foresters and some gardeners. Food items consumed in winter are bark, needles and buds of conifers. Bark, twig tips and buds of alder, willow and aspen also are favored.

Most rabbits and hares are essentially nonvocal as adults, but snowshoe hares master several calls. They are capable of producing an ear-piercing distress cry when injured or captured. Imitation of this scream is used by hunters to attract predators such as coyotes. A grunting sound expresses the animal's anger or fear. Thumping their hind feet to alarm others of impending danger, or as a part of the courtship dance is yet another element in their acoustic repertoire.

Populations exhibit a peculiar phenomenon not yet fully understood by scientists. Their numbers may fluctuate from almost none to extreme over-population. The interval between population peaks is about nine or 10 years, with the cycles most pronounced in the northern part of their range in Alaska or Canada. Suggested causes of the fluctuations have included changes in the food supply, or contagious diseases that accompany overcrowding. The "stress syndrome" leads to fighting and retarded reproduction and is thought to be triggered by high population numbers. Predators, rather than causing the cycles, seem merely to tag along for the ride. When hare populations crash, so do those of the predators that depend on them as a major food source.

The snowshoe hare is the primary food source for the predatory lynx, each one of which consumes about 200 of the hares per year. Alan Carey photo.

Pygmy Rabbit and Mountain Cottontail

Almost every farm boy growing up in the Midwest, Northeast or South has taken more than a passing interest in rabbits at one time or another. Rabbit hunting is almost one of the rites of passage of youth in many parts of the country. In the Northwest, however, the humble rabbit takes a decided back seat to more glamorous and substantial mule deer, elk or bighorn sheep. Due to lack of hunter interest, state wildlife departments largely ignore the rabbit as a game animal in the Northwest, and those who breed beagle hounds don't have much of a market for their wares.

Unless you raise a garden, rabbits are unobtrusive, even desirable, local residents. Of those native to the New World, the grayish pygmy rabbit (*Sylvilagus idahoensis*) is the smallest, averaging only 10 inches long and weighing one-half to one pound. The grayish-brown mountain cottontail (*Sylvilagus nuttalli*) grows somewhat bigger, with an average size of 13 inches and a weight of about two pounds. A pygmy rabbit's tail is gray, while the cottontail's is a pure-white puff. Pygmies have tiny whitish spots on their cheeks on either side of the nose; the special marking of the mountain cottontail is its black-edged ears.

Pygmy rabbits' range in the Northwest is mostly restricted to southern Idaho and eastern Oregon, with an isolated, legally protected population in the semi-arid plains of east-central Washington. They prefer sagebrush flats interspersed with rabbitbrush, greasewood and bunch grasses. Here in the loosely packed soil, they excavate extensive burrow systems, unlike any other rabbits in North America. A naturalist with the Northern Railroad Survey of 1853-55 probably unknowingly referred to the pygmy rabbit when he reported, "I was told of another kind of small rabbit of a bluish tint, shorter ears, and which burrowed in the ground, but I could not get any." Other rabbits, like the mountain cottontail, prefer using burrows left by other animals.

The mountain, or Nuttall's, cottontail lives in a variety of habitats, with mountains probably being the least preferred. Generally, they select brushy thickets along borders between forest and

Nuttall's cottontail, Audubon lithograph. Audubon, Quadrupeds. *Photo from Special Collections Division, University of Washington Libraries.*

On Stone by Wm E. Hitchcock

grassland, but they also frequent brush-covered rock outcrops or sagebrush thickets in more open areas. This cottontail also may be found in the shelterbelts between grainfields, showing its capability of coexisting with man as it expands its range as a result of farming activities. Brush piles from field clearing offer them the cover necessary for resting during the day and the convenience of nearby food supplies. They have even been found to "hole up" under abandoned machinery on many farms.

Both rabbits are most active during the late evening and early morning hours. Pygmy rabbits spend their days in burrows, while shallow depressions under dense vegetation or a rocky crevice provide shelter for the cottontail. Neither animal ranges far from its home area. The pygmy stays, in general, within 30 yards of its burrow complex; the home range of a mountain cottontail seldom exceeds an acre.

Pygmy rabbits have a rather monotonous diet of sagebrush both summer and winter, while cottontails seek a summer cuisine of grasses and succulent plants, spiced with sage. In winter more woody vegetation, such as twig tips and bark appear on the cottontail's menu.

Both species breed in spring and summer. An average of three litters, each with five or six bunnies, arrives approximately one month apart during summer, enabling a female to produce about two-dozen offspring a year. This legendary reproductive capability has a purpose, since rabbits and their kin act as one of the most important links in the food chain between plant material and the various carnivores. Hungry enemies abound, including most birds of prey, foxes, coyotes, weasels, dogs, cats and even snakes. There are few, if any, meat-eating animals that would decline a meal of rabbit.

Marsupials

Opossum. Tom and Pat Leeson photo.

Opossum

Suburban residents near Portland or Seattle might be surprised to know that a relative of the Australian koala "bear" and kangaroo may frequent their backyards. Small hand-shaped footprints near garbage cans in an alley mark the path of the only American marsupial, the opossum (*Didelphis virginianus*). Possums carry and feed their tiny young in a belly-mounted pouch, or marsupium, in much the same way as kangaroos.

The family to which opossums belong is the New World's oldest and most primitive, with a fossil record going back more than 100 million years. The opossum has made it through at least the past 20,000 years without noticeable changes, showing its tremendous biological adaptability.

The possum is one of the few creatures whose range has *increased* since the arrival of the white man in America. Originally they were found only in the southeastern corner of the United States, as suggested by their species name, *virginianus*. Their arrival in the Northwest has been linked to the thousands of Appalachian highlanders who emmigrated to western Washington's Lewis-Cowlitz and Skagit-Snohomish counties between 1890 and 1920. These hill folk recreated their Appalachian lifestyle in the mountain recesses of the Cascades and Willapa Hills, bringing with them traditional plants, including the black walnut and catnip, and their prized opossum, considered a delicacy in the traditional Appalachian diet. Today in the Northwest, opossums are most common from the seashore east to the Cascades, but range expansion also has occurred up the Columbia River, and an isolated population in eastern Idaho has spread into western Montana.

Over its Northwest range, this versatile animal may be found in nearly any habitat from sea level to 5,000-foot-high mountains, but it seems to do best when near humans, possibly because food can be more easily scavenged. Bill Meyer, Regional Public Affairs Officer with the U.S. Fish and Wildlife Service Regional Office in Portland, notes that opossums are so numerous in the Portland area that the telling of "Possum Jokes" has become a favored local pastime.

Although widespread within their range, possums seldom are seen because of their nocturnal lifestyle. Tom and Pat Leeson photo.

Opossums are omnivorous, feeding on a wide variety of fruits, seeds, forbs and even carrion. They are especially fond of young birds and mammals, but will even eat poisonous snakes, to whose vemon they usually are immune. Opossums appear to grow throughout their short, one-year lives, reaching about the size of a housecat. The face of this scruffy gray mammal is white with a pointed snout and naked black ears tipped with white. The tail is long (10 to 21 inches), scaly and prehensile, meaning it can grasp objects. It is used, among other things, for climbing, balancing, carrying nest materials and hanging from a tree limb.

The possum builds its nest in hollow trees, brush piles, another animal's burrow, under houses and even in suburban garages. The nest is lined with dry grasses and leaves, which are picked by mouth, and then securely grasped by the tail. The whole mass is then trundled along behind as the animal makes its way to the nest.

Primarily nocturnal, they are seldom seen by humans except in the headlights of an automobile or dead on the road. Possums are basically antisocial, leading solitary lives except during the mating season, when battles over females can be furious. Their 13-day gestation period is the shortest of any North American animal. At birth the navy bean-size young are blind and naked, and greatly resemble embryos. Immediately upon birth, the 20-plus newborns must make an arduous three-inch journey to nourishment at the more than dozen nipples in the pouch. Young are weaned at 75 to 100 days of age, but remain with the mother up to four months, emerging then for frequent piggyback rides.

The opossum is an accomplished actor worthy of a Wild Kingdom Oscar. Its first reaction to an antagonist is hissing, drooling and showing off its 50 gleaming teeth, more than any other North American land mammal can boast. If this display fails, it feigns death, or "plays possum," assuming a cataleptic state resembling temporary paralysis. Motionless on its side with partly closed eyes, it lolls its tongue and drools. The "play" may last from a few minutes up to six hours.

Its most dreaded enemy moves on four wheels rather than four legs. Other more traditional predators include great-horned owls, dogs and, in some places, those rare people who consider possum pie a delicacy.

Opossums traditionally have been considered simple-minded, but recent laboratory experiments suggest otherwise. In several tests possums beat such "smart" animals as dogs, cats and rabbits in remembering where food was hidden in a complicated maze. In fact, only humans appeared to be absolutely superior to the little creature in such studies! Regardless of how dumb he may or may not be, the lowly opossum helps to demonstrate an important lesson. A high degree of evolutionary specialization and/or high intelligence is not always correlated with high success in survival over the long run. We have witnessed the magnificent, highly specialized bighorn sheep decline to a few isolated remnant populations, while the ancient opossum is gaining ground every year. Maybe this is a lesson for much younger modern man to ponder.

Extremely adaptable animals, opossums have flourished in the Northwest, even to the point of becoming pests in urban areas. Tom and Pat Leeson photo.

"Will opossums bring prosperity to Toledo?"

The opossum may be one of America's most adaptable creatures. Although only a short-time resident of the Northwest, this introduced species is making its presence known, even to city dwellers. From major metro areas to small towns, possums have moved in and set up housekeeping. In Portland, poking fun at the less-than-beautiful invaders now is in vogue. For many regional residents, including Donna Marsh Allocca of Centralia, Washington, the neighborhood marsupials still are a novelty. "We have one in the garage and put dog food out for him," she reports. Twenty miles to the south at Toledo, the "possum infestation" has been a media event.

Dateline Toledo, Washington
February 15, 1985

For the past week the hottest news here has revolved around a furry little creature described as the "ugliest doggone bugger you've ever seen."

It's an opossum, or possum as they are affectionately called in these parts. And that ugly little bugger has done more for this city of 600 in the past week than everyone else has done in the past year.

THE ATTENTION IT brought Toledo is phenomenal. In addition to stories in The Daily Chronicle, this "opossum invasion" has drawn reporters or calls from The Oregonian of Portland, The Seattle Times, The Spokesman-Review of Spokane, The Bellingham Herald and The Longview Daily News. Many of the stories were front page features.

Mayor Shirley Grubb was even interviewed by reporters from Portland's KATU-TV. The interview also aired on KOMO-TV in Seattle.

Then there are the calls Grubb received from ABC and NBC radio services in New York.

No one gets any sleep here because opossums keep them awake with their constant grunting and groaning. They live in woodpiles, in basements, under porches and in attics.

THE MAIL BROUGHT Grubb numerous other solutions. A lady from Kirkland wrote to suggest the mayor rid the opossums by putting crushed moth balls in her attic, a remedy she claims also works on raccoons.

A lady in Vader sent a four-page letter describing in detail how to build a box trap.

A group of sixth graders at Winlock Miller Elementary school sent the mayor several suggestions, including establishing a bounty, getting more street lights so drivers will be able to see more and thus get more road kills, and importing Japanese Ninja experts to fling their deadly metal stars at the opossums.

BUT, THE ULTIMATE solution to the opossum problem is probably one found in a followup media letter, to be sent out this week by Wade and the Chamber of Commerce.

Wade said the critters should be shipped to garbage facilities around the country to act as live garbage disposals.

"By giving possums first crack at the city's garbage and knowing their breeding potential the tonnage of possum meat to be harvested annually is staggering beyond belief," the letter reads.

"A by-product of this trade would be possum pelts that we find New York city officials are definitely interested in for their use as a crime deterrent. Coats would be made of possum hides and sold to New Yorkers. Upon the approach of a mugger, they would fall down, play dead, and after awhile the mugger gets tired of waiting around and leaves in disgust."

Excerpts courtesy of *The Daily Chronicle*, Centralia-Chehalis, Washington.

Epilogue

Despite more than a century and a half of permanent white settlement, the Northwest still is home to a great number and variety of wild mammals. This rich heritage must not be taken for granted. Ongoing changes already have taken their toll on wildlife and its irreplaceable habitat. Continuing population growth, economic expansion, environmental pollution and even more leisure time and greater disposable incomes will mean continuing threats to our wild Northwest residents.

Today, more than ever, an interested and knowledgeable citizen has the ability to affect important decisions, since public input now is an integral part of the planning process. Fortunately, Northwesterners have not been afraid to speak out on national forest plans, national park plans, game management proposals and other issues when they feel the interest of wildlife is being compromised.

Northwesterners who want to know more about wild mammals and become involved with helping to assure that our wildlife heritage can be passed to future generations, might consider joining one of many effective conservation organizations, from small local groups to large and powerful national associations. Among a long list of highly regarded national organizations, many with local or state affiliates are:

National Wildlife Federation
1412 16th Street, N.W.
Washington, D.C. 20036

Ranger Rick's Nature Club
1412 16th Street, N.W.
Washington, D.C. 20036
(Children's Division of the
National Wildlife Federation)

National Audubon Society
950 Third Avenue
New York, New York 10022

Whale-watcher off the Oregon Coast delights in this touching moment as she pets a young gray whale. Eva Cooley photo.

Defenders of Wildlife
1244 19th Street, N.W.
Washington, D.C. 20036

American Cetacean Society
P.O. Box 4416
San Pedro, CA 90732

Further Reading

R. D. Burroughs, ed., *The Natural History of the Lewis and Clark Expedition* (East Lansing: Michigan State University Press, 1961), 340p.

E. E. Clark, *Indian Legends of the Pacific Northwest* (Berkeley: University of California Press, 1953), 225p.

D. E. Gaskin, *The Ecology of Whales and Dolphins* (London: Heinemann Publishing Company, 1982), 459p.

L. G. Ingles, *Mammals of the Pacific States—California, Oregon and Washington* (Stanford: Stanford University Press, 1965), 506p.

E. B. Kritzman, *Little Mammals of the Pacific Northwest* (Seattle: Pacific Search Press, 1977), 120p.

E. J. Larrison, *Mammals of the Northwest—Washington, Oregon, Idaho and British Columbia* (Seattle: Seattle Audubon Society, 1976), 256p.

A. Savage and C. S. Savage, *Wild Mammals of Northwest America* (Baltimore: Johns Hopkins University Press, 1981), 209p.

Bighorn sheep. Richard E. Kirchner photo.

The Photographers

The following professional nature/wildlife photographers contributed most of the contemporary mammal pictures in this volume. Color prints of these and other wildlife subjects can be ordered directly from the photographers.

Alan Carey
P.O. Box 118
Lolo, MT 59847

Thomas W. Kitchin
62 East 21st Avenue
Vancouver, British Columbia
Canada V5V 1P5

Bruce Pitcher
P.O. Box 1381
Bozeman, MT 59715

Richard E. Kirchner
P.O. Box 1261
Bozeman, MT 59715

Ken Balcomb, III
1359 Smuggler's Cove
Friday Harbor, WA 98250

Tom & Pat Leeson
P.O. Box 2498
Vancouver, WA 98668

Tom Spaulding
P.O. Box 383, Pike Place
Seattle, WA 98101

Northwest Geographer™ Series

No. 1, The inaugural volume

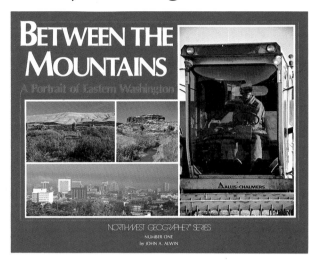

BETWEEN THE MOUNTAINS: A Portrait of Eastern Washington, by John A. Alwin
ISBN 0-9613787-0-0

In *BETWEEN THE MOUNTAINS* geographer John Alwin takes you through Washington's big, open country east of the Cascades and introduces you to the land and people of this unique region. Already a classic, the 128-page book includes almost 250 pictures in brilliant NORTHWEST GEOGRAPHER color, two dozen historic photos, maps and a 50,000-word, non-technical text.

The reviews are in:

"fascinating text is enhanced by 248 color photos . . . The book promises great things for the series."
Seattle Times/Seattle Post Intelligencer
Seattle, WA

"beautiful color photographs . . . alone make them [the books] a bargain . . . text is livelier than a reader might expect."
Northwest: The Oregonian's Sunday Magazine
Portland, OR

"an informative and affectionate overview of the region and its people."
The Spokesman-Review
Spokane, WA

"slick, lavishly illustrated softbound book is neither a text nor a coffee-table pictorial. It is, instead, an informational showcase of Eastern Washington written by a still-curious expert in the field."
Yakima Herald-Republic
Yakima, WA

John A. Alwin, a university geography professor for ten years, is now editor and publisher of the NORTHWEST GEOGRAPHER Series. He logged more than 10,000 miles, interviewed scores of residents and took more than 5,000 slides researching for *BETWEEN THE MOUNTAINS.*

Next in the Series, No. 3

SEATTLE-TACOMA AND THE SOUTHERN SOUND
by Ronald R. Boyce

Why did Seattle rise to its present prominence over other cities in Puget Sound? How does the past help us understand the region today? Why is the Puget Sound area growing so rapidly? What is truly distinctive about Seattle and sister cities on the Sound?

To find out join Ronald R. Boyce on his latest urban expedition, calculated to capture the special character of Seattle and other cities on the southern Sound. Lucidly written, with more than 150 beautiful color photographs and dozens of historic pictures, this book takes you back in time, lets you walk today's waterfronts, wonder at the rise of skyscrapers, sort out the suburbs and explore the water world of the Sound. From Everett to Olympia and Bremerton to Bellevue, it's all here in words and pictures.

Come along with Ronald Boyce in *SEATTLE-TACOMA AND THE SOUTHERN SOUND,* Number 3 in the NORTHWEST GEOGRAPHER Series, and discover a place you thought you knew.

Ronald R. Boyce is an internationally known urban geographer with more than 20 books to his credit. He was a professor of geography at the University of Washington for more than a decade and currently is Dean, School of Social and Behavioral Sciences at Seattle Pacific University. A long-time resident, Boyce knows the cities of the Sound well.

Don't miss a single volume. To complete your collection, or to be added to the Northwest Geographer Series mailing list and receive information on pre-publication discounts on forthcoming books, write:

Northwest Panorama Publishing, Inc.
NORTHWEST GEOGRAPHER Series
P.O. Box 1858
Bozeman, MT 59771